AF595030

Sensor Networks in Structural Health Monitoring

Sensor Networks in Structural Health Monitoring

From Theory to Practice

Editors

Eleni Chatzi
Vasilis K. Dertimanis

MDPI • Basel • Beijing • Wuhan • Barcelona • Belgrade • Manchester • Tokyo • Cluj • Tianjin

Editors
Eleni Chatzi
ETH Zürich
Switzerland

Vasilis K. Dertimanis
ETH Zürich
Switzerland

Editorial Office
MDPI
St. Alban-Anlage 66
4052 Basel, Switzerland

This is a reprint of articles from the Special Issue published online in the open access journal *Journal of Sensor and Actuator Networks* (ISSN 2224-2708) (available at: https://www.mdpi.com/journal/jsan/special_issues/sensor_networks_structural_health_monitoring).

For citation purposes, cite each article independently as indicated on the article page online and as indicated below:

LastName, A.A.; LastName, B.B.; LastName, C.C. Article Title. *Journal Name* **Year**, *Volume Number*, Page Range.

ISBN 978-3-0365-0632-6 (Hbk)
ISBN 978-3-0365-0633-3 (PDF)

Contents

About the Editors

Eleni Chatzi is currently an Associate Professor and the Chair of Structural Mechanics at the Institute of Structural Engineering of the Department of Civil, Environmental and Geomatic Engineering (DBAUG) of ETH Zürich. She obtained her diploma (2004) and MSc (2006) in Civil Engineering, with honors, from the Department of Civil Engineering at the National Technical University of Athens (NTUA). In June 2010, she obtained her Ph.D. with distinction from the Department of Civil Engineering & Engineering Mechanics at Columbia University. In 2010, she was hired as the youngest Assistant Professor at ETH, and she was promoted to an Associate Professor in 2017.

Prof. Chatzi's research couples novel simulation tools with state-of-the-art monitoring methodologies for intelligent and data-driven assessment and diagnostics of engineered systems, with the goal of providing actionable tools able to guide operators and engineers in the management of their assets. A key aspect of her research lies in the extraction of quantifiable metrics that are indicative of structural performance across the component, system, and network levels. Her expertise lies in the area of structural health monitoring, with a strong focus on problems lying beyond the commonly adopted assumption of linear time-invariant systems. Her research spans a broad range of topics, including applications in emerging sensor technologies and structural control, methods for curbing uncertainties in structural diagnostics and life-cycle assessment, and advanced schemes for nonlinear/nonstationary dynamics simulations.

Prof. Chatzi serves as editor for numerous peer-reviewed international journals with particular focus on system identification methods and topics relating to SHM. Since 2016, she has coordinated the joint ETH Zürich & University of Zurich Ph.D. Programme in Computational Science. She is currently leading the ERC Starting Grant WINDMIL on the topic of "Smart Monitoring, Inspection and Life-Cycle Assessment of Wind Turbines", awarded by the European Research Council. Her work in the domain of self-aware infrastructure was recognized with the 2020 Walter L. Huber Research Prize, awarded by the American Society of Civil Engineers (ASCE). She is also a recipient of the 2020 EASD Junior Research Prize in the area of Computational Structural Dynamics, awarded by the European Association of Structural Dynamics (EASD).

Vasilis K. Dertimanis, Senior Assistant & Lecturer: Vasilis was born in Greece. He received his diploma in Mechanical Engineering from the University of Patras, Greece, and his Ph.D. from the National Technical University of Athens (NTUA), Greece, in the area of modeling and identification of faults in mechanical and structural systems. His research interests lie in the areas of structural identification and health monitoring, linear and nonlinear state estimation, active and passive structural control, and hybrid testing and optimization.

Vasilis has served as a senior researcher in the NTUA Vehicles Laboratory, Machine Design Laboratory, and Laboratory for Earthquake Engineering. He has also participated as a Marie Curie experienced researcher in the EU-funded SmartEN ITN project. For more than a decade, he has been in parallel self-employed as a freelance engineer and inspector, as an instructor in training seminars on transportation of dangerous goods by road/rail, and as a measurement engineer and structural vibration analyst.

Since January 2014, Vasilis has been a member of the Chair of the Structural Mechanics at ETH Zurich, and since May 2017, he has been Senior Assistant (Oberassistent) actively supporting the Chair in research and teaching.

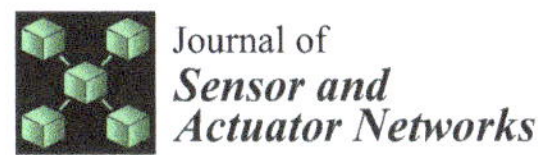
Journal of
Sensor and
Actuator Networks

Editorial

Sensor Networks in Structural Health Monitoring: From Theory to Practice

Vasilis Dertimanis * and Eleni Chatzi

Department of Civil, Environmental and Geomatic Engineering, ETH Zürich, Stefano-Franscini-Platz 5, 8093 Zürich, Switzerland; chatzi@ibk.baug.ethz.ch
* Correspondence: v.derti@ibk.baug.ethz.ch

Received: 5 October 2020; Accepted: 12 October 2020; Published: 13 October 2020

The growing attention that structural health monitoring (SHM) has enjoyed in recent years can be attributed, amongst other factors, to the advent of low-cost and easily deployable sensors. The enabling technology has brought forth a new era of structural diagnostic means and is continuously redefining the tools for information processing, data reduction/compression, feature extraction, and smart assessment.

Within the SHM community, novel data-driven or hybrid methods are being developed, demonstrated on field deployments on large-scale systems, including transport, wind energy, and building infrastructure, promoting the actionability of SHM as an essential resource for life-cycle and resilience management. Nonetheless, in optimizing these deployments, a number of open issues remain with respect to the sensing side. These are associated with the type, configuration, and eventual processing of the information acquired from these sensors in order to deliver continuous behavioral signatures of the monitored structures.

Within this context, the aim of this Special Issue is to discuss the latest advances in the field of sensor networks for SHM. The focus lies in both active research on the theoretical foundations of optimally deploying and operating sensor networks, as well as in those technological developments that might designate the next generation of sensing solutions targeted for SHM.

Initiating form the data interpretation side, Argyris et al. [1] introduce a Bayesian framework for finite element model updating using experimental modal data. The novelty lies in utilization of a new likelihood formulation, which enables inclusion of the modal shapes, based on probabilistic treatment of the modal amplitude coherence (MAC) value between the model-predicted and measured mode shapes. The actual case study of the Metsovo bridge is utilized for validation. The modal characteristics of the bridge are identified based on operational modal analysis methods by means of a network of reference and roving sensors. The transitional Markov chain Monte Carlo algorithm is employed for model updating. The method delivers robust uncertainty bounds, outperforming conventional Bayesian formulations, which result in unrealistically thin uncertainty bounds for the model parameters, when a large (redundant) number of sensors is employed.

Moving away from the Bayesian paradigm, Pai et al. [2] describe a framework for practical asset management. In their work, the authors try to relax strong assumptions on uncertainty placed by typical data-interpretation methods, such as Bayesian model updating and residual minimization to better address practical engineering challenges. Error-domain model falsification is compared against these two more established approaches in terms of the ability to provide robust, accurate, engineer-friendly, and computationally inexpensive results for engineering model updating. Via demonstration on two full-scale case studies, the authors argue that error-domain model falsification is able to incorporate, iteratively and transparently, the incremental information gain, stemming from the sensing networks to facilitate updating of structural models at low additional computational cost.

Moving to utilization of sensing for triggering control actions, Azimi and Molaei Yeznabad [3] exploit a wireless-sensor network (WSN) as a main enabler for seismic vibration mitigation. Inspired by swarm intelligence, they propose a new control method, termed swarm-based parallel control (SPC), to improve seismic performance and minimize pounding hazards. This is achieved via the WSN-enabled exchange of information among adjacent buildings at corresponding floor levels. Three-dimensional time histories are simulated under earthquake excitation, with the response mitigated via use of semi-actively controlled magnetorheological (MR) dampers. The proposed control algorithm is assessed against fuzzy logic control (FLC), as well as passive on/off methods. Results indicate that in case of failure in the control system, as well as structural damage, the proposed WSN-driven SPC method can sense damage and accordingly update the control forces in the adjacent buildings to alleviate pounding.

A key factor in successful modal identification, model updating, and control is the chosen configuration of sensors. In this special issue, Barthorpe and Worden [4] offer a review of advances in sensor placement optimization (SPO) strategies for structural health monitoring (SHM). In recent years, this has been primarily examined in the context of damage identification. This work formally defines the SPO for SHM problem, offers an overview of the current state of the art, and proceeds to identify promising emergent trends. Some of the major (still) open questions pertain to robustness to modelling uncertainty, benign measurement influences, and sensor failures. A lack of sufficient studies on practical validation of the developed sensor designs is further remarked. It is finally argued that, since such a design is carried out at an offline design stage, it is not required to devote more effort in accelerating the proposed optimization schemes.

The issue of sensing extends beyond what is deemed as traditional civil infrastructure. In their work, Al-Obaidi et al. [5] study the transport processes of solids due to geophysical flows via use of a novel sensing mechanism; a miniaturized instrumented particle that can provide a direct, non-intrusive, low-cost, and accessible method for the assessment of coarse sediment particle entrainment. The presented instrumented particle is fitted with a triaxial MEMs accelerometer, a magnetometer and angular displacement sensors, which enable monitoring of the particle's 3D motion at a configurable rate of 200–1000 Hz. An appropriate filter is used for inertial sensor fusion, uncertainty mitigation, and noise reduction. The instrumented particle is tested under controlled lab conditions, reproducing initiation of the destabilization of a bed surface in an open channel flow. The work showcases the potential of custom designed and appropriately calibrated instrumented particles in the context of monitoring the possible instigation of water infrastructure hazards.

The final contribution underlines the importance of delivering methods able to tackle the challenge of uncertainty in the environmental and operational conditions (EOCs) that a monitored system is exposed to, and more importantly, do so in absence of availability of physics-based models. In a conventional setting, varying dynamic response can be captured via use of parameter-varying extensions of the original autoregressive with exogenous input (ARX) model. Yet, EOCs tend to vary in time-scales (days, weeks, months) that are sufficiently longer to the system dynamics. Tatsis et al. [6] propose a distributed framework for structural identification, damage detection, and localization by extending the conventional ARX structure. Here, the long time-scale dependence on EOCs is captured via a Gaussian process regression, while a distributed residual generation algorithm is set up for damage detection. Finally, damage localization is demonstrated via computation of the ARX-inferred mode shape curvatures. The diagnostic capabilities of the proposed scheme under varying EOCS are verified on two simulated case studies of increasing complexity.

This special issue has proposed contributions across the complete SHM information chain; from sensor design to configuration (placement), data interpretation, and triggering of reactive action (control). The featured papers offer an overview of the state-of-the-art and further proceed to introduce novel methods and tools. Several of these particularly target the treatment of uncertainty, which inherently describes the sensed information and the behavior of monitored systems. It is hoped that the readers will appreciate this Special Issue enhancing their knowledge in the area and helping them advance their ongoing research and innovation activities in sensing for SHM.

Conflicts of Interest: The authors declare no conflict of interest.

References

1. Argyris, C.; Papadimitriou, C.; Panetsos, P.; Tsopelas, P. Bayesian model-updating using features of modal data: Application to the Metsovo Bridge. *J. Sens. Actuator Netw.* **2020**, *9*, 27. [CrossRef]
2. Pai, S.; Reuland, Y.; Smith, I. Data-interpretation methodologies for practical asset-management. *J. Sens. Actuator Netw.* **2019**, *8*, 36. [CrossRef]
3. Azimi, M.; Molaei Yeznabad, A. Swarm-based parallel control of adjacent irregular buildings considering soil–structure interaction. *J. Sens. Actuator Netw.* **2020**, *9*, 18. [CrossRef]
4. Barthorpe, R.; Worden, K. Emerging trends in optimal structural health monitoring system design: From sensor placement to system evaluation. *J. Sens. Actuator Netw.* **2020**, *9*, 31. [CrossRef]
5. Al-Obaidi, K.; Xu, Y.; Valyrakis, M. The design and calibration of instrumented particles for assessing water infrastructure hazards. *J. Sens. Actuator Netw.* **2020**, *9*, 36. [CrossRef]
6. Tatsis, K.; Dertimanis, V.; Ou, Y.; Chatzi, E. GP-ARX-Based structural damage detection and localization under varying environmental conditions. *J. Sens. Actuator Netw.* **2020**, *9*, 41. [CrossRef]

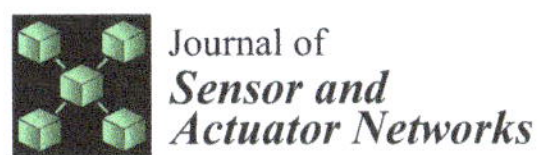
Journal of
Sensor and
Actuator Networks

Article

Bayesian Model-Updating Using Features of Modal Data: Application to the Metsovo Bridge

Costas Argyris [1,†], Costas Papadimitriou [1,*], Panagiotis Panetsos [2] and Panos Tsopelas [3]

[1] Department of Mechanical Engineering, University of Thessaly, 38334 Volos, Greece; costasargyris@gmail.com
[2] Capital Maintenance Department, Egnatia Odos S.A., 57001 Thermi, Greece; panpanetsos@gmail.com
[3] Department of Mechanics, School of Applied Mathematical and Physical Sciences, National Technical University of Athens, 15773 Athens, Greece; tsopelas@central.ntua.gr
* Correspondence: costasp@uth.gr
† Current position: Software Engineer, The MathWorks, Business Park, Cambridge CB4 0HH, UK.

Received: 31 January 2020; Accepted: 29 May 2020; Published: 3 June 2020

Abstract: A Bayesian framework is presented for finite element model-updating using experimental modal data. A novel likelihood formulation is proposed regarding the inclusion of the mode shapes, based on a probabilistic treatment of the MAC value between the model predicted and experimental mode shapes. The framework is demonstrated by performing model-updating for the Metsovo bridge using a reduced high-fidelity finite element model. Experimental modal identification methods are used in order to extract the modal characteristics of the bridge from ambient acceleration time histories obtained from field measurements exploiting a network of reference and roving sensors. The Transitional Markov Chain Monte Carlo algorithm is used to perform the model updating by drawing samples from the posterior distribution of the model parameters. The proposed framework yields reasonable uncertainty bounds for the model parameters, insensitive to the redundant information contained in the measured data due to closely spaced sensors. In contrast, conventional Bayesian formulations which use probabilistic models to characterize the components of the discrepancy vector between the measured and model-predicted mode shapes result in unrealistically thin uncertainty bounds for the model parameters for a large number of sensors.

Keywords: Bayesian inference; model updating; modal identification; structural dynamics; bridges

---

1. Introduction

The evaluation of the actual dynamic characteristics of structures, such as modal frequencies, modal damping ratios and mode shapes, through vibration measurements, as well as the development of high-fidelity finite element (FE) models, has been attracting an increasing research effort worldwide. Measured response data of structures mainly under ambient vibrations offer an opportunity to study quantitatively and qualitatively their dynamic behavior. These vibration measurements can be used for estimating the modal properties of structures, as well as for updating the corresponding FE models used to simulate their behavior [1,2]. The information for the calibrated FE models and their associated uncertainties is useful for checking design assumptions, for validating the assumptions used in model development, for improving modeling and exploring the adequacy of the different classes of FE models, and for carrying out more accurate robust predictions of structural response. These models are representative of the initial structural condition of the structure and can be further used for structural health-monitoring purposes [3–7].

Bayesian methods for ambient (operational) modal identification [8–18] and structural model updating [19–31] are used to develop high fidelity FE models of structures using modal properties identified from ambient vibration measurements. Due to the large size of civil infrastructure, the mode shapes are assembled from a number of sensor configurations that include optimally-placed reference sensors as well as moving sensors [32]. The modal properties are then integrated within Bayesian model-updating formulations to calibrate the parameters of large-scale FE models, as well as their associated uncertainty. The goal is to develop accurate and reliable models of the actual structures that are proven to closely simulate their behavior.

As far as the computational part is concerned, for complex posterior distributions, stochastic simulation algorithms such as Transitional Markov Chain Monte Carlo (TMCMC) [33] can be conveniently used to sample from the posterior distribution for parameter estimation, model selection and uncertainty propagation purposes. These methods require a large number of forward model runs which can increase the computational effort to excessive levels if one simulation for a high-fidelity large-order FE model requires several minutes or even hours to complete. For that purpose, fast and accurate component mode synthesis (CMS) techniques, consistent with the FE model parameterization [34,35], are used to achieve drastic reductions in computational effort. Further computational savings are achieved by adopting a parallelized version of the TMCMC algorithm to efficiently distribute the computations in available multi-core CPUs [36,37].

A novel likelihood function formulation is introduced in this work, which treats mode shapes not as full vectors, but as scalars using features between the measured and model-predicted mode shapes such as the MAC value. Instead of following the conventional Bayesian approach of assigning a multivariable Gaussian distribution to the error vector quantifying the discrepancy between the measured and model predicted mode shapes, a truncated Gaussian distribution is proposed for the probabilistic modeling of the scalar MAC value between the model predicted and experimental mode shapes. This effectively reduces the number of data points in the likelihood and leads to different uncertainty quantification results compared to the classic vector-based likelihood formulation. It is demonstrated that the proposed formulation has certain desired properties which can not be obtained under the vector-based formulation for the likelihood.

The capabilities of the proposed modal-based Bayesian model-updating methodology are demonstrated by calibrating the parameters of a high-fidelity FE model developed for the Metsovo bridge, using modal properties experimentally identified from ambient vibration data. The FE model is parametrized with respect to the stiffnesses of the deck, piers and soil components of the bridge. Ambient acceleration time histories from multiple points along the bridge deck are used to extract the modal properties of the bridge experimentally, and the identified modal properties are used as data in the Bayesian model updating methodologies in order to perform inference about the model parameters. In order to explore the effect of soil–structure interaction, two classes of models are examined and compared using Bayesian model selection [26,38]. Comparisons between the vector-based and the proposed MAC-based likelihood formulations demonstrate the advantages of the MAC-based likelihood formulation.

This work is structured as follows. Section 2 presents the Bayesian inference framework for FE model parameter estimation using modal properties. Section 2.1.1 reviews existing likelihood formulations, while Section 2.1.2 present the new formulation for building the likelihood based on features between experimental data and model predictions. The use of model reduction techniques to alleviate the computational burden encountered with sampling techniques is summarized in Section 2.2. Section 2.3 briefly outlines the whole procedure of parameter estimation and uncertainty propagation using the TMCMC sampler. The field structure is introduced in Section 3, along with the unreduced and reduced FE models of the structure, and the experimental modal identification procedure. Section 4 presents the results of model updating based on the experimentally-identified modes and demonstrates the advantages of the proposed MAC-based likelihood formulation. Conclusions are summarized in Section 5.

2. Bayesian Parameter Estimation Using Modal Data

To apply the Bayesian formulation for parameter estimation of linear FE models, we consider that the data D consists of the squares of the modal frequencies, $\hat{\omega}_r^2$, and the mode shapes $\underline{\hat{\phi}}_r \in R^{N_{0,r}}$, $r = 1, \ldots, m$, experimentally estimated using vibration measurements, where m is the number of identified modes and $N_{0,r}$ is the number of measured mode shape components for mode r. Consider also a parameterized linear FE model class $\tilde{M}$ of a structure and let $\underline{\theta} \in R^{N_\theta}$ be a vector of free structural model parameters to be estimated using the set of modal properties identified from vibration measurements.

Let $\omega_r(\underline{\theta})$ and $\underline{\phi}_r(\underline{\theta}) \in R^{N_{0,r}}$ be the r-th modal frequency and mode shape at $N_{0,r}$ measured DOFs, respectively, predicted by the model for a given value $\underline{\theta}$ of the model parameters. The squares of the modal frequencies $\omega_r^2(\underline{\theta})$ and the mode shape components $\underline{\phi}_r(\underline{\theta}) = L_r \underline{\varphi}_r(\underline{\theta}) \in \mathbb{R}^{N_{0,r}}$ are computed from the full mode shapes $\underline{\varphi}_r(\underline{\theta}) \in \mathbb{R}^n$ that satisfy the eigenvalue problem:

$$[K(\underline{\theta}) - \omega_r^2(\underline{\theta}) M(\underline{\theta})] \underline{\varphi}_r(\underline{\theta}) = \underline{0}, \tag{1}$$

where $K(\underline{\theta}) \in \mathbb{R}^{n \times n}$ and $M(\underline{\theta}) \in \mathbb{R}^{n \times n}$ are the global stiffness and mass matrices respectively of the FE model of the structure, n is the number of model DOFs, and $L_r \in R^{N_{0,r} \times n}$ is an observation matrix, usually comprised of zeros and ones, that maps the n model DOFs to the $N_{0,r}$ observed DOFs for mode r. For a model with large number of DOFs, $N_{0,r} \ll n$.

The likelihood $p(D|\underline{\theta}, \tilde{M})$ is the probability of observing the measured data D under the model $\tilde{M}$ for parameters equal to $\underline{\theta}$. It is used in Bayes rule to update the posterior distribution $p(\underline{\theta}|D, \tilde{M})$ of the model parameters $\underline{\theta}$ as follows:

$$p(\underline{\theta}|D, \tilde{M}) = \frac{p(D|\underline{\theta}, \tilde{M})\, p(\underline{\theta}|\tilde{M})}{p(D|\tilde{M})}, \tag{2}$$

where $p(\underline{\theta}|\tilde{M})$ is the prior distribution of the model parameters and $p(D|\tilde{M})$ is the evidence of the model class, selected so that $p(\underline{\theta}|D, \tilde{M})$ integrates to one.

2.1. Likelihood Formulation

The likelihood formulation is of critical importance in Bayesian inference. To build the likelihood, one needs to assume a probabilistic relation between the model predictions and experimental data in order to account for unavoidable model error as well as experimental or measurement error. There is not just one way to do that, and different likelihood formulations can lead to different results. Therefore, Bayesian inference is subjective in the sense that different likelihood models can be tried using the same data, and the inference results might differ significantly. Prediction error equations, which relate the model predictions with the experimental data probabilistically, are used to formulate the likelihood. Depending on the nature of the data, different prediction error equations can be used for different subsets of the entire data set.

For the modal frequencies, the most common choice is the uncorrelated Gaussian error assumption for each modal frequency (e.g., [39,40]). Specifically, the prediction error equation for the r-th modal frequency is taken as:

$$\hat{\omega}_r^2 = \omega_r^2(\underline{\theta}) + \varepsilon_{\omega_r}, \tag{3}$$

where ε_{ω_r} is the prediction error for the r-th modal frequency taken to be Gaussian with zero mean and standard deviation $\sigma_{\omega_r} \hat{\omega}_r^2$. The unknown parameter σ_{ω_r} is included in the parameter set $\underline{\theta}$ to be estimated from the data. This formulation for the modal frequencies assumes that each modal frequency is uncorrelated with the rest. Then, the likelihood term for the r-th modal frequency is the probability of observing the measured frequency given specific values of the model parameters $\underline{\theta}$, derived from Equation (3) in the form:

$$p(\hat{\omega}_r^2|\underline{\theta}) = N(\hat{\omega}_r^2; \omega_r^2(\underline{\theta}), \sigma_{\omega_r}^2 \hat{\omega}_r^4), \tag{4}$$

where $N(x;\mu,\sigma^2)$ denotes the univariate Gaussian PDF evaluated at point x with mean μ and variance σ^2.

However, as far as the mode shapes are concerned, the prediction error formulation can be more complex due to the fact that they are vectors with multiple components. Again we make the assumption that all mode shapes are uncorrelated with each other and therefore we can treat each mode shape individually, just like the modal frequencies. Two formulations are presented next. The first one is a review of existing formulations, while the second one is a novel formulation based on features between model predicted and experimentally identified mode shapes.

2.1.1. Formulation Using Probabilistic Models for Mode Shape Vectors

An often-used formulation for the prediction error is to assume that the discrepancy vector between the measured mode shape vector and the model predicted mode shape vector follows a zero-mean multivariable Gaussian distribution with a specified covariance matrix. The prediction error equation for the r-th mode shape is then

$$\hat{\underline{\phi}}_r = \beta_r(\underline{\theta})\underline{\phi}_r(\underline{\theta}) + \varepsilon_{\underline{\phi}_r}, \tag{5}$$

where $\varepsilon_{\underline{\phi}_r}$ is the prediction error vector for the r-th mode shape taken to be Gaussian with zero mean and covariance matrix $\sigma_{\underline{\phi}_r}^2 \Sigma_{\underline{\phi}_r}$, where the matrix $\Sigma_{\underline{\phi}_r}$ specifies the possible correlation structure between the components of the prediction error vector of the r-th mode shape, the unknown scalar $\sigma_{\underline{\phi}_r}^2$ is included in the parameter set to be estimated, and

$$\beta_r(\underline{\theta}) = \hat{\underline{\phi}}_r^T \underline{\phi}_r(\underline{\theta}) \,/\, \left\|\underline{\phi}_r(\underline{\theta})\right\|^2 \tag{6}$$

is a normalization constant such that the measured mode shape $\hat{\underline{\phi}}_r$ at the $N_{0,r}$ measured DOFs is closest to the model mode shape $\beta_r(\underline{\theta})\underline{\phi}_r(\underline{\theta})$ predicted by the particular value of $\underline{\theta}$, and $||\underline{z}||^2 = \underline{z}^T\underline{z}$ is the usual Euclidean norm. The scalar $\beta_r(\underline{\theta})$ is introduced in Equation (6) to account for the fact that the measured modeshape $\hat{\underline{\phi}}_r$ is normalized to have Euclidean norm equal to one, while the model predicted modeshape $\underline{\phi}_r(\underline{\theta})$ is mass normalized. The scalar $\beta_r(\underline{\theta})$ is derived by minimizing the distance $||\hat{\underline{\phi}}_r - \beta_r(\underline{\theta})\underline{\phi}_r(\underline{\theta})||$ between the measured mode shape and the scaled version of the model predicted mode shape.

It is important to note in this approach that the number of data points used for each mode shape is equal to the number of measured DOFs $N_{0,r}$ for that particular mode. For a spatially uncorrelated model for the prediction error $\varepsilon_{\underline{\phi}_r}$ (diagonal $\Sigma_{\underline{\phi}_r}$ matrix) each mode shape component counts as a new independent data point in the likelihood. From the Bayesian Central Limit Theorem, the posterior uncertainty is expected to reduce without bounds as the number of mode shape components is increased. However, as the number of measured DOFs increases, the sensors become very close to one another, providing almost the same information content that should not further reduce the posterior uncertainty of the model parameters. The closeness of the sensors depends on the wavelength of the considered measured mode shape. Two sensors are close and are expected to provide redundant information if their distance is a fraction of the wave length of the corresponding mode shape. Therefore, a spatially uncorrelated model for the prediction error vector $\varepsilon_{\underline{\phi}_r}$ of the mode shape would not yield the expected behavior regarding posterior uncertainty as the number of mode shape components increases.

A remedy to this is to introduce a correlation model between the components of the prediction error vector of the mode shape, leading to a non-diagonal covariance matrix $\Sigma_{\underline{\phi}_r}$. However, a correlation function should be postulated to describe the spatial correlation between two mode shape components (sensors) as a function of their distance, where the closer two sensors get the more they are correlated. Several correlation functions exist in the literature [41]. The problem is that one cannot know beforehand which correlation function is the proper one for the particular application at hand. This decision of the correlation function might turn out to be extremely difficult to make in practice, because in practical situations one normally has slight to none available information regarding the correlation nature of the prediction error vector. Selecting the proper correlation function might be challenging and failure to do so could easily lead to erroneous results as was demonstrated in [41]. Finding the proper correlation function is not the goal of this work. More on that issue can be found in [41–43]. Herein two cases of correlation models are examined: uncorrelated and exponentially correlated models.

For the simplest case of uncorrelated mode shape prediction error vectors the covariance matrix simplifies to a diagonal matrix:

$$\Sigma_{\underline{\phi}_r} = ||\hat{\underline{\phi}}_r||^2 / N_{0,r}\, I, \tag{7}$$

with I being the $N_{0,r} \times N_{0,r}$ identity matrix, while for the exponentially correlated model the identity matrix I is replaced by the correlation matrix R_r whose (i,j)-th element is given by the exponential correlation function:

$$R_r(i,j) = \exp(-x_r(i,j)/\lambda_r), \tag{8}$$

where $x_r(i,j)$ is the Euclidean distance between the i-th and j-th mode shape components (sensors) for the r-th mode, and λ_r is the correlation length for the r-th mode which is a parameter to be identified.

Using Equation (5), the likelihood term for the r-th mode shape is the probability of observing the measured mode shape for given model parameters $\underline{\theta}$, given by

$$p(\hat{\underline{\phi}}_r|\underline{\theta}) = N(\hat{\underline{\phi}}_r; \beta_r(\underline{\theta})\underline{\phi}_r(\underline{\theta}), \sigma^2_{\underline{\phi}_r}\Sigma_{\underline{\phi}_r}(\underline{\theta})), \tag{9}$$

where $N(\underline{x}; \underline{\mu}, \Sigma)$ denotes the multivariate Gaussian PDF evaluated at point $\underline{x}$ with mean vector $\underline{\mu}$ and covariance matrix Σ. Following the work of Papadimitriou et al. [44] which was based on the same prediction error Equation (5), the likelihood function in Equation (9) can be expressed in terms of the MAC values between the measured and model predicted mode shapes.

Slightly different prediction error equations for the mode shapes have been proposed in the literature (e.g., [39,40]), including versions that do not require the use of the mode correspondence [4,6]. In all these alternatives, the likelihood formulation for the mode shapes is based on a probabilistic description of individual components of a vector and thus they fall into the category discussed in this subsection.

2.1.2. Formulation Using Probabilistic Models for MAC Values

The previous formulation uses the mode shapes as full vectors in the likelihood function. Herein we propose a novel formulation for including the mode shapes in the likelihood function which is based on the MAC value between the experimental and model predicted mode shape. The MAC value, defined as $MAC(\underline{u}, \underline{v}) = \underline{u}^T\underline{v}/(||\underline{u}||\;||\underline{v}||)$ between two vectors $\underline{u}$ and $\underline{v}$, is the most common way to measure the similarity between two mode shape vectors. It is a scalar measure which varies from 0 to 1 with a value of 1 indicating a perfect match. The scaling of the mode shapes is not important for the MAC value which means that no normalization is needed for either the experimental or model predicted mode shape.

In the new formulation the experimental mode shape is not compared with the model predicted mode shape in an element-wise fashion, but rather based on its MAC value. This reduces the number of data points used in the likelihood for each mode shape from $N_{0,r}$ to just 1. Therefore, instead of calculating the probability of observing the experimental mode shape vector given the model predicted mode shape vector (for some given model parameter values), we calculate the probability of their MAC value taking a value of 1, implying that they match perfectly.

In contrast to the previous vector formulation of the likelihood, the MAC value is a univariate quantity and therefore requires a univariate distribution to model it. Taking into account the fact that the MAC value is strictly bounded in the interval [0, 1], a Truncated Gaussian distribution is used, although there are many other choices of candidate distributions. The Gaussian distribution is preferred because of its known properties. This leads to the following prediction error equation for the MAC value of the r-th mode shape:

$$\hat{MAC}_r = MAC_r(\underline{\theta}) + \varepsilon_{MAC_r}, \tag{10}$$

where $MAC_r(\underline{\theta}) = MAC(\underline{\hat{\phi}}_r, \underline{\phi}_r(\underline{\theta}))$ is the model-predicted MAC value, defined as the MAC value between the experimental r-th mode shape and the model predicted r-th mode shape for the given values of the model parameters $\underline{\theta}$. The term ε_{MAC_r} is the error in the r-th MAC value (analogous to the error in the r-th frequency), assumed to follow a univariate zero-mean Gaussian distribution with standard deviation equal to σ_{MAC_r}. The standard deviation σ_{MAC_r} is a measure of "how far" the observed MAC value $\hat{MAC}_r$ can be from the model-predicted MAC value $MAC_r(\underline{\theta})$ due to model and experimental errors. This can be thought of as completely analogous to the error term for the modal frequencies in (3). The resulting Gaussian with mean $MAC_r(\underline{\theta})$ and standard deviation σ_{MAC_r} is truncated in 0 and 1 which results in the Truncated Gaussian distribution.

An important issue that should be addressed when using MAC values is the fact that although the MAC is a scalar value, it depends on the number of mode shape components used. This needs to be taken into account in the formulation in order to avoid erroneous results. For example, if only two components of a mode shape are used, there is a chance that the MAC value turns out to be very close to 1 (provided that those two components match well between the two mode shapes). However, if a large number of components is used, due to small errors in each component there is the chance that the MAC value is significantly lower than 1, which would mean that the case with two components would yield a larger MAC value. However, the case of large number of components components is expected to be much more informative than the case of two components since the more components we have the better we know the actual geometry of the mode shape. This naturally leads to the conclusion that the number of mode shape components must be taken into account, assigning higher preference to MAC values calculated with more components than MAC values calculated with less components.

One way to account for this in a Bayesian framework is through manipulation of the MAC value standard deviation parameter σ_{MAC_r}. We seek a formula through which to define σ_{MAC_r} that depends on the number of mode shape components $N_{0,r}$. Although there is not only one way to achieve this, the following formula is used:

$$\sigma_{MAC_r} = \sigma'_{MAC_r}\left(1 + \frac{1}{\sqrt{N_{0,r}}}\right), \tag{11}$$

where σ'_{MAC_r} is the parameter to be inferred from data. The first term in Equation (11) describes the uncertainty present in the MAC value that exists independently of the number of sensors. This uncertainty exists even for a large number of sensors and is due to model and experimental errors in the individual components and can not be reduced further. The second term in Equation (11) depends on the number of sensors and gets smaller as the number of sensors is increased, which reduces the standard deviation of the MAC value. This way more weight (less uncertainty) is given to MAC values calculated with more sensors. These are modeling choices within the Bayesian framework, much like the choice of Gaussian PDFs for the likelihood, independent data, etc. Alternative formulations could also be postulated. In particular, the two terms in Equation (11) can be weighted differently but this falls outside the scope of the present work.

Then the likelihood term for the MAC value of the r-th mode shape is the probability of observing a MAC value of 1 for given values of the model parameters $\underline{\theta}$ (indicating a perfect match between the experimental and model predicted mode shapes), given by the Truncated Gaussian PDF:

$$p(M\hat{A}C_r = 1|\underline{\theta}) = TN(1; MAC_r(\underline{\theta}), \sigma^2_{MAC_r}, 0, 1), \tag{12}$$

where $TN(x; \mu, \sigma^2, a, b)$ denotes the Truncated Gaussian PDF evaluated at point x with mean μ, variance σ^2 and truncation limits a and b.

2.1.3. Likelihood Formulation Combining Modal Frequencies and Mode Shapes

The parameter set $\underline{\theta}$ of the structural model class $\tilde{M}$ is augmented to include the parameters σ_{ω_r} and $\sigma_{\underline{\phi}_r}$ or σ'_{MAC_r} related to the prediction error models. For simplicity, in order to avoid having too many parameters, the three prediction error parameters are assumed to be the same for all modes and therefore their dependence on r is dropped.

The total likelihood function is easily calculated as the product of the individual likelihoods for the frequencies and mode shapes, given their independence. For the vector formulation of the mode shapes the likelihood is:

$$p(D|\underline{\theta}, \tilde{M}) = \prod_{r=1}^{m} p(\hat{\omega}_r^2|\underline{\theta}) \prod_{r=1}^{m} p(\hat{\underline{\phi}}_r|\underline{\theta}), \tag{13}$$

where $p(\hat{\omega}_r^2|\underline{\theta})$ and $p(\hat{\underline{\phi}}_r|\underline{\theta})$ are given by (4) and (9) respectively. For the MAC formulation of the mode shapes the likelihood is:

$$p(D|\underline{\theta}, \tilde{M}) = \prod_{r=1}^{m} p(\hat{\omega}_r^2|\underline{\theta}) \prod_{r=1}^{m} p(M\hat{A}C_r = 1|\underline{\theta}) \tag{14}$$

where $p(\hat{\omega}_r^2|\underline{\theta})$ and $p(M\hat{A}C_r = 1|\underline{\theta})$ are given by (4) and (12), respectively.

2.2. *Computational Tools*

The transitional Markov chain Monte Carlo algorithm (TMCMC) [33] is used for estimating the parameters of FE models by drawing samples from the posterior probability density function of the model parameters. Markov chain Monte Carlo algorithms, including TMCMC used in this work, require a moderate to very large number of repeated system analyses to be performed over the space of uncertain parameters. Consequently, the computational demands depend highly on the number of system analyses and the time required for performing a system analysis. For FE models with large number of DOFs, this can increase substantially the computational effort to excessive levels. Computational savings are achieved by adopting parallel computing algorithms to efficiently distribute the computations in available multi-core CPUs [36,37,45].

In addition, fast and accurate CMS techniques [46], consistent with the finite element model parameterization, are integrated with Bayesian techniques to reduce efficiently and drastically the FE model and thus reduce the computational effort [34,35]. CMS techniques are widely used to analyze structures in a reduced space of generalized coordinates. CMS involves dividing the structure into a number of substructures (components), obtaining reduced-order models of the substructures keeping a fraction of the substructure modes, and then assembling a reduced order model for the entire structure using the kept substructure modes and interface degrees of freedom between substructures. Additional substantial reductions can be achieved by reducing the number of interface DOF using characteristic interface modes through a Ritz coordinate transformation [34]. However, for methods involving re-analyses due to variations in the values of the uncertain model parameters the reduction for computing the system modes has to be repeated for each re-analysis. This gives rise to a substantial computational overhead that arises from the model reduction at component level, and from assembling the component mass and stiffness matrices to form the reduced global system mass and stiffness matrices. The main objective in methods involving re-analyses of models with varying properties is to completely avoid the re-analysis at the component level as well as the re-assembling of the reduced global matrices at the system level.

It has been shown that when the partition of the structure into substructures is guided by certain parameterization schemes, the reduced global mass and stiffness matrices derived using CMS techniques can be represented exactly by an expansion of these matrices in terms of scalar functions of the model parameters, with coefficient matrices computed and assembled once from a single CMS analysis of a reference structure [34,47,48]. This representation allows one to re-compute the reduced global stiffness and mass matrices for different values of the model parameters from these expansions, avoiding expensive re-analyses involved in CMS procedure. Dramatic reduction in computational effort has been reported without compromising the accuracy in the modal properties predicted by the reduced model.

The reduction achieved by applying the CMS technique in the FE model of the Metsovo bridge is described in Section 3.3.

2.3. Outline of Procedure

Given the parameterized FE model of a structure, a parameterized reduced FE model is first obtained using CMS. This amounts to forming the reduced global stiffness and mass matrices as a function of the model parameters $\underline{\theta}$. The TMCMC sampler was used to sample from the posterior PDF in Equation (2), where the likelihood function is given either by Equations (4), (9) and (13) for the vector-based formulation or by Equations (4), (12) and (14) for the MAC-based formulation. The modal properties involved in the likelihood function are computed for each TMCMC sample in the model parameter space using the reduced FE model. Specifically, for each one of the two likelihood formulations presented in Sections 2.1.1 and 2.1.2, the reduced stiffness and mass matrices are used in Equation (1) to predict the modal properties $\omega_r(\underline{\theta})$ and $\underline{\phi}_r(\underline{\theta})$ for different values of the model parameter set $\underline{\theta}$. The sample points $\underline{\theta}^{(j)}, j = 1, \ldots, N$, obtained from the TMCMC sampler populate the posterior PDF of the model parameters. These samples are subsequently used to depict the uncertainties in the model parameters and propagate uncertainties in output Quantity of Interest (QoI) by providing estimates of the modal frequencies $\omega_r(\underline{\theta})$ and MAC values $MAC(\hat{\underline{\phi}}_r, \underline{\phi}_r(\underline{\theta})), j = 1, \ldots, N$, using Equation (1) for the reduced FE model. Results of uncertainty quantification are expressed in terms of marginal distributions for the model parameters, as well as useful simplified measures of uncertainty, such as mean and credible intervals of the output QoI.

3. Application to Metsovo Bridge

3.1. Description of Bridge

The ravine bridge of Metsovo (Anthohori–Anilio tunnel) of Egnatia Motorway is crossing the deep ravine of Metsovitikos river, 150 m over the riverbed. A picture of the bridge is shown in Figure 1. This is the highest bridge of the Egnatia Motorway, with the height of the tallest pier equal to 110 m. The total length of the bridge is 537 m. The bridge has 4 spans of length 44.78 m, 117.87 m, 235 m, 140 m and 3 piers of which the closest to the left side of the bridge (Figure 1), with height of 45 m, supports the boxbeam superstructure through pot bearings (movable in both horizontal directions), while the other two piers, with heights 110 m and 35 m, connect monolithically to the structure.

3.2. Finite Element Model of Bridge

The detailed geometry of the bridge is complicated because the deck and the piers have variable cross-sections and the deck is also inclined. A high fidelity FE model of the bridge is created using three-dimensional tetrahedral quadratic Lagrange finite elements. The model takes into account the potential soil-structure interaction by modeling the soil with large blocks of material and embedding the piers and abutments into these blocks. The nominal values of the moduli of elasticity of the deck and piers are selected to be the values used in design: 37 GPa for the deck and 34 GPa for the piers. The nominal value of the soil is taken to be 1 GPa. The largest size of the elements in the mesh is of the order of the thickness of the hollow deck cross-section. The size of the FE mesh is chosen to predict the first 20 modal frequencies and mode shapes of the bridge with sufficient accuracy. Several mesh sizes were tried, and an accuracy analysis was performed in order to find a reasonable trade-off between the number of degrees of freedom (DOF) of the model and the accuracy in the predicted modal frequencies. A mesh of 830,115 DOFs was kept for the bridge-soil model. This mesh was found to cause errors of the order of 0.1–0.5% in the first 20 modal frequencies, compared to the smallest possible mesh sizes which had approximately 3 million DOFs.

Figure 1. Picture of the Metsovo bridge.

The intent is to build a high fidelity model that could, in future studies, be extended locally to incorporate nonlinear mechanisms activated during strong motion or deterioration phenomena. In this study the focus is to update a baseline linear model using low-intensity vibration measurements. In future studies, the availability of higher-intensity vibration measurements will provide data for improving modeling and updating parameters of nonlinear models introduced to represent localized nonlinear phenomena activated due to large vibrations or deterioration due to various damage mechanisms. Simplified beam models, although adequate for design purposes, are inadequate to use for setting up digital twins of structures so that are reliable under various operating conditions. Simplified modeling, for example with beam elements, does not offer an adequate representation of the system dynamics over the dynamic range activated by various operational conditions. Such simplified models are often inadequate for monitoring purposes and involve large model errors even for operational conditions under which the structure may be assumed to behave linearly.

3.3. Model Reduction Using CMS

The time required for a complete run of the FE model is approximately 2 min on a 8-core 3.20 GHz computer. Due to the thousands of forward model runs for different values of the model parameters that are required by the Bayesian computational tools, it is necessary to reduce the time required for a single model run. Model reduction is used to reduce the model size and thus the computational effort to manageable levels. Specifically, the parameterization-consistent CMS technique [34,35] based on the Graig-Bampton method [46] is applied to the bridge-soil FE model.

For this, the bridge is divided into 16 physical components with 15 interfaces between the components. Specifically, the deck is divided into six components or substructures of length 120 m, 120 m, 60 m, 50 m, 117 m and 70 m each. One component is assigned to each one of the three piers. Two components are introduced for the left and right abutments of the bridge. Five more components are introduced for the large solid blocks representing the flexibility of the soil at the connections with the three piers and the two abutments. This partition into component is one of the many alternative ones, introduced herein to demonstrate the capabilities of CMS technique for model reduction. Usually the partition of the structure into components is guided by the purpose of the analysis or the structural health monitoring goals. For example, components may be introduced to monitor and select models of nonlinearities activated by various operational conditions in isolated (localized) parts of a structure. The partition of a structure into components facilitates monitoring of the structural health, allowing the identification of the location and severity of sparse damage within a small subset of substructures.

For each component, it is selected to retain all modes that have frequency less than $\omega_{\max} = \rho\omega_c$, where $\omega_c = 3.52$ Hz is the cut-off frequency selected to be equal to the 20th modal frequency of the nominal FE model. The ρ values affect the computational efficiency and accuracy of the CMS technique. For $\rho = 5$ selected for most components, a total of 170 internal DOFs out of the 814,080 are retained for all 16 components. The total number of DOFs of the reduced model is 16,205 which also includes 16,035 interface DOFs. It is clear that more than an order of magnitude reduction in the number of DOFs is achieved using CMS. The largest fractional error between the modal frequencies computed using the complete FE model and the ones computed using the CMS technique for $\rho = 5$ falls below 0.2%. Thus a very good accuracy is achieved.

The large number of the interface DOFs can be reduced by retaining only a fraction of the constrained interface modes [34,49]. For each interface, only the modes that have frequency less than $\omega_{\max} = \nu\omega_c$ are retained, where ν is user and problem dependent. For $\nu = 200$ selected for most interfaces, the largest fractional error for the lowest 20 modes of the structure falls below 0.43%. In particular, for $\nu = 200$ and $\rho = 5$ the reduced system has 1891 DOFs from which 170 generalized coordinates are fixed-interface modes for all components and the rest 1721 generalized coordinates are constrained interface modes [34]. A trade-off was made between reducing the model as much as possible (fewer kept DOFs) and keeping the accuracy of the predicted modal frequencies as close as possible to those of the unreduced model. It should be noted that further reductions are possible using an enhanced substructuring technique where the dynamics contribution of several kept modes is replaced by their static contribution [47].

Thus, using CMS a drastic reduction in the number of DOFs is obtained which can exceed two orders of magnitude, without sacrificing the accuracy with which the lowest 20 modal frequencies are computed. The time to solution for one run of the reduced model is of the order of a few seconds which should be compared to approximately 2 min required for solving the unreduced FE model.

Moreover, for nonlinear models of structures, especially models where local nonlinearities are mainly activated, the model reduction techniques can also be applied to reduce the models of components of the structure that behave linearly under various operational conditions [35,48].

3.4. Experimental Modal Identification

The testing system consist of a movable array of servo-accelerometers that are usually being installed on the bridge deck (sidewalks or pavement surface) or inside the box beam internal voids to measure the vibrations (accelerations) of the bridge under ambient excitations. The available measurement system consisted of five triaxial and three uniaxial accelerometers paired with a 24-bit data recording system, a GPS module for synchronization between sensors, and a battery pack. The system is wireless and can be easily moved from one location in the structure to another. The recorder can connect with a laptop through wired (Ethernet) or wireless (Wi-Fi) connection to be set up in the desired way (sampling rate, recording duration, repeater recordings etc) or view the measurements while they are being recorded for quality checking. Given the limited number of sensors and the large length of the deck, the entire length of the deck was covered in 13 sensor configurations, shown in Figure 2. For each configuration the recording lasted 20 min at a sampling rate of 100 Hz. Each triaxial sensor was positioned on the bridge sidewalks such that it measures along the transverse, vertical and longitudinal directions of the bridge deck. One triaxial and three uniaxial sensors (one vertical and two horizontal transverse) remained in the same position throughout the measurements, in order to provide common measurement points amongst different configurations such as to enable the assembling of the total mode shape from partial mode shape components measured from the different configurations [30,32]. The use of more than one reference sensors per direction guarantees the redundancy of the measuring scheme in case one sensor is placed at the node of the modeshape. The wireless feature of the measurement system allows the execution of all recordings over the 13 sensor configurations in a single day. The recorded responses are mainly due to road traffic, which ranged from light vehicles to heavy trucks, and environmental excitation such as wind loading, which classifies this case as ambient (operational) modal identification.

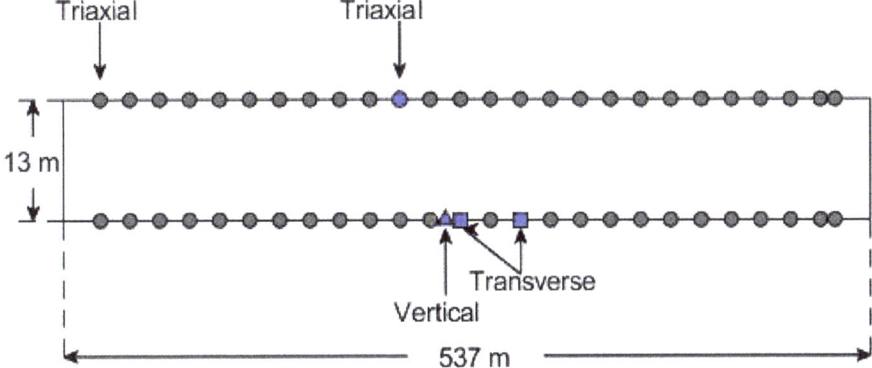

Figure 2. Measured locations in the bridge. Solid grey circles represent the locations of the triaxial sensors measuring along the longitudinal, transverse and vertical direction of the bridge deck. Solid blue circle shows the location of the reference triaxial sensor. Solid blue rectangle denotes the locations of the two uniaxial sensors measuring along the transverse direction. Solid blue triangle denotes the location of uniaxial sensor measuring along the vertical direction.

The Bayesian operational modal analysis methodology [9,10] is used to estimate the modal frequencies, mode shapes and damping ratios for each sensor configuration. The mode shapes are assembled from the local mode shapes of each configuration using the methodology proposed by Au [32]. The full mode shapes are produced at all 159 sensor locations covered by the 13 sensor configurations. The components along the longitudinal direction of the bridge deck are ignored. Only the components along the transverse and vertical direction of the bridge deck are processed. The output-only vibration measurement for some of the 13 sensor configurations were not reliable enough to estimate the mode shape components at higher modes. As a result, it was not possible to assemble the mode shapes for more than 12 modes. Thus these mode shapes were excluded from the analysis. Specifically, the first 20 modal frequencies and modal damping ratios of the bridge were identified, along with 11 mode shapes. The mode shapes of all the modes up to the 12th were identified, except the 10th mode which was very poorly identified and also excluded from the date set. Table 1 presents the mean and standard deviation of the experimentally identified modal frequencies for all 20 identified modes of the Metsovo bridge. It also compares the identified frequencies and mode shapes with those predicted by the nominal FE model. In particular, the experimental and nominal model predicted mode shapes are compared using their MAC value which is a scalar measure of correlation between two mode shapes ranging from 0 to 1, with a value of 1 indicating perfect correlation. The identified mode shapes are shown in Figures A1–A4 of Appendix A and compared with the corresponding mode shapes predicted by the nominal FE model of the bridge. From both the MAC values of Table 1 and mode shapes of Figures A1–A4 it can be clearly seen that the mode shapes predicted by the nominal FE model match very accurately the corresponding experimentally identified mode shapes with MAC values higher than 0.95 for the 11 identified mode shapes (except mode 9 which has a MAC value of 0.87). However, there appears to be a significant mismatch between the experimental and nominal FE model modal frequencies which indicates that a finite element model updating should be performed in order to achieve a closer fit between the model predicted and the experimentally identified modal frequencies.

Table 1. First 20 experimentally identified modal frequencies of the the Metsovo bridge compared with the modal frequencies predicted by the nominal FE model of the bridge. MAC values between identified and model predicted mode shapes are also shown.

Mode	Experimental Frequency (Hz)	Nominal Model Frequency (Hz)	Mode Shape MAC Value
1	0.306	0.293	0.991
2	0.603	0.574	0.955
3	0.623	0.619	0.978
4	0.965	0.849	0.935
5	1.047	1.050	0.969
6	1.139	1.070	0.965
7	1.428	1.388	0.972
8	1.697	1.578	0.966
9	2.005	1.690	0.878
10	2.303	1.966	0.85
11	2.367	2.156	0.972
12	2.590	2.317	0.971
13	2.723	2.500	
14	3.086	2.745	
15	3.127	2.815	
16	3.480	2.876	
17	3.861	2.950	
18	4.058	3.320	
19	4.210	3.381	
20	4.410	3.521	

4. Model Updating Results

The FE model of the bridge–soil system is parameterized using three parameters associated with the modulus of elasticity of the deck (θ_1), piers (θ_2) and soil (θ_3). The model parameters multiply the nominal values of the corresponding moduli of elasticity for the deck (37 GPa), the piers (34 GPa) and the soil (1 GPa). The nominal values for the deck and piers are reasonable estimates since they are the moduli of elasticity of the concrete used in design and therefore their updated values of θ_1 and θ_2 are expected to lie close to 1. However, as far as the soil is concerned, its nominal value is only a rough estimate, based on soil property measurements conducted at the site of the bridge. Therefore, its nominal value should be dealt with a large uncertainty in the model updating procedure. These modeling considerations regarding the initial parameter uncertainties are taken into account in the Bayesian framework through the prior PDF. It should be noted that a simplified uniform parameterization involving a small number of parameters is considered in order to avoid possible unidentifiability issues and enable the comparison between the two different likelihood formulations.

4.1. Model Updating Using Modal Frequencies Only

First, the FE model of the bridge-soil system is updated using only a subset of the experimentally identified modal frequencies. This approach allows one to use the rest of the frequencies in order to validate the updated model by checking its predictive capabilities with data that was not used in the updating. Specifically, the first 15 identified modal frequencies are used to estimate the model parameters and their uncertainty, while the other five modal frequencies are used in order to validate the updated model. For 11 out of the 20 modes we use mode correspondence through the MAC values to associate the experimentally identified and model predicted modal properties. It was found that the i-th experimentally identified mode corresponds to the i-th mode predicted by the model. For modes higher than 12 for which there no mode shape identified from experimental data, we match the modal frequencies based only on the number of mode identified or predicted by the FE model, with modal frequencies arranged in an ascending order.

4.1.1. Flexible-Soil Model

The prior distribution for the parameters are assumed to be uniform with bounds in the domain $[0.5, 1.5] \times [0.5, 1.5] \times [0.1, 1000]$ for the deck, pier and soil parameters respectively, and in the domain $[0.001, 1]$ for the prediction error parameter σ_ω. The domain for the soil parameter was deliberately chosen much larger in order to account for the large uncertainty in the values of the soil stiffness and be able to explore the full effect of the soil stiffness on the model behavior.

Model updating results are obtained using the parallelized TMCMC algorithm [33,36] for the bridge–soil FE model. The TMCMC is used to generate samples from the posterior PDF of the structural model and prediction error parameters. These samples represent the posterior PDF and therefore our updated state of knowledge about the parameters given the experimental data. After the posterior samples are drawn the parameter uncertainty is propagated to the predictions of the first 20 modal frequencies of the bridge. This is done in order to check the fit of the updated model with the experimental frequencies that were used to perform the model updating, but also with the next five modal frequencies that were not included in the data set. In all TMCMC runs, the following selection is made: $TolCov = 1.0$, $\beta = 0.2$ [33]. The number of samples used per TMCMC stage are 1000, resulting in a total runtime of approximately 10 minutes using the reduced 1.891 DOF model in a 8-core 3.20 GHz computer.

The TMCMC samples which represent the posterior PDF are visualized through their marginal distributions and two-dimensional (2D) projections in Figure 3. The sample statistics are shown in Table 2. The posterior parameter uncertainty is propagated through the model using the samples to yield the robust model predictions of the lowest 20 modal frequencies. The fit is shown in Figure 4. The predicted modal frequencies are normalized with respect to the experimentally-identified frequencies for comparison convenience. Therefore, values close to 1 are close to the experimental frequencies. The improvement achieved by the updated model compared to the nominal model is evident. For most modes the experimental frequency lies within the predicted 5–95% interval or very close to it, and in all cases the error is of the order of 3–4% which should be compared to the error of the nominal model which is of the order of 10% to 20% for some modes. This is a strong indication of the need for model updating in order to improve the accuracy and predictive capability of the updated model.

Regarding the parameters, it can be seen that the updated values of the deck and pier parameters lie close to 1 as expected, and slightly below it. The mean values for the deck and pier stiffness parameters are estimated to be approximately 0.95 and 0.98 times their nominal values with uncertainties of the order of 5% and 12% respectively. From the $(\theta_1 - \theta_2)$ 2D projection of Figure 3 it is evident that a negative correlation exists between the deck and pier stiffnesses. This is reasonable since an increase in the stiffness of the deck can be counterbalanced by a decrease in the stiffness of the piers such that the modal frequency values are maintained, and vice versa.

As far as the updated soil stiffness is concerned, the only (but important) new information that is acquired by the model updating is that its value can be arbitrarily large, as long as it exceeds a threshold. The threshold value appears to be approximately 70 which is the minimum value that the updated soil parameter can attain, as seen from its posterior marginal distribution in Figure 3. A value of 70 implies a soil modulus of elasticity of 70 GPa which is more than double of the updated (and nominal) value of the pier modulus of elasticity (34 GPa). The soil parameter can increase substantially above this value without affecting the fit with the experimental data, that is, without causing any variation in the predicted modal frequencies of the model. Considering that the uniform prior bound for the soil stiffness was set to $[0.1, 1000]$ it is obvious that lower values which would attribute to the soil some flexibility similar to that of the piers are not preferred. In addition, the large posterior uncertainty in soil property indicates that the modal frequencies are insensitive to the values of the soil modulus of elasticity for these high values of the soil property. This insensitivity is due to the low vibration levels recording from ambient operational conditions of the bridge.

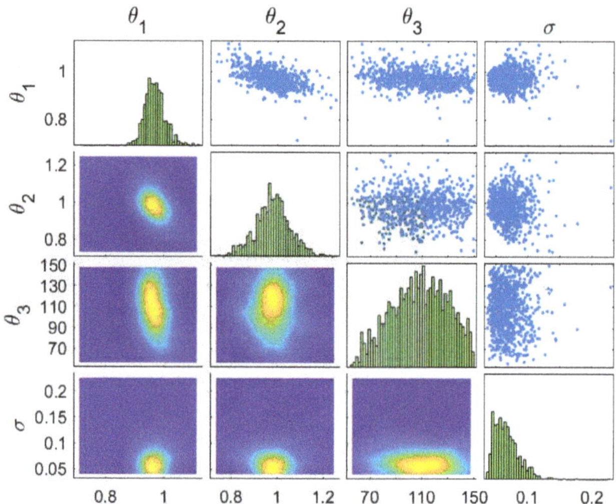

Figure 3. Posterior marginal distributions and 2D sample projections of model parameters. θ_1: deck, θ_2: piers, θ_3: soil, σ: prediction error, log-evidence = 2.527.

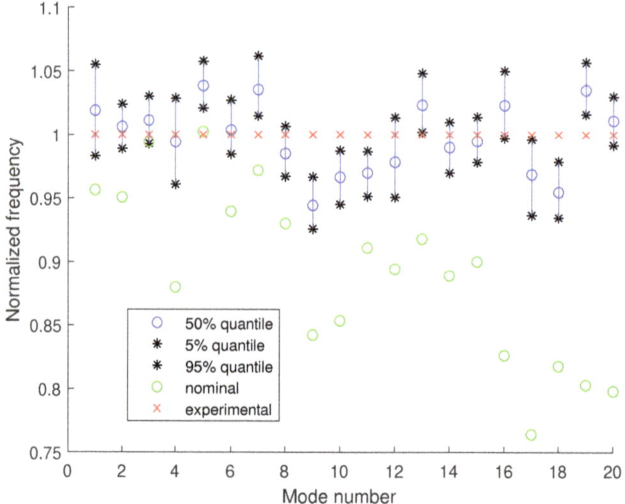

Figure 4. Uncertainty propagation to the first 20 modal frequencies compared with the experimental frequencies and nominal model predictions.

Table 2. Quantiles of posterior samples for all parameters. The x% quantile for a parameter is defined as the level that corresponds to x% of the values of the parameter over all TMCMC samples, arranged in increasing order, to fall below this level.

	θ_1	θ_2	θ_3	σ
5% Quantile	0.919	0.844	70.428	0.044
50% Quantile	0.968	0.981	110.23	0.065
95% Quantile	1.029	1.115	150.20	0.105

4.1.2. Two-Parameter Stiff-Soil Model

The results obtained from the flexible-soil model suggest that the bridge appears to be fixed to the ground and the modal properties predicted by the model are insensitive to the soil modulus of elasticity. This leads to introducing a second model, which corresponds to eliminating the soil parameter by fixing its value to a large value as suggested by its posterior marginal distribution of Figure 3, simulating the very stiff soil conditions which were found from the first model. Therefore, the new two-parameter model has as parameters the modulus of elasticity of the deck (θ_1) and piers (θ_2), while the soil parameter is fixed to 100.

The posterior samples for the two-parameter model are visualized using their marginal distributions and 2D projections in Figure 5. The sample statistics are shown in Table 3. The posterior parameter uncertainty is propagated through the model using the samples to yield the robust model predictions of the lowest 20 modal frequencies. The fit is shown in Figure 6. Note that in Figure 6 the predictions of the nominal model are closer to the experimental due to the increase of the soil parameter to the fixed value of 100 in order to simulate the stiff-soil conditions, which led to an increase of the modal frequencies of the nominal model. It can be seen that, as expected, the model updating results both in terms of the updated values of the parameters and in terms of the fit with the data are almost identical to the results obtained from the three-parameter model in Figure 4.

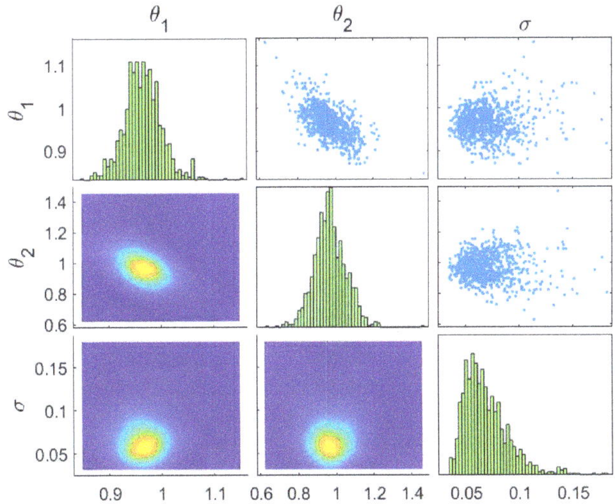

Figure 5. Posterior marginal distributions and 2D sample projections of model parameters. θ_1: deck, θ_2: piers, σ: prediction error, log-evidence = 2.550.

Table 3. Quantiles of posterior samples for all parameters.

	θ_1	θ_2	σ
5% Quantile	0.908	0.835	0.044
50% Quantile	0.965	0.966	0.065
95% Quantile	1.022	1.108	0.108

This is also confirmed using the Bayesian model selection framework [38] to compute the evidence $p(D|M_i)$ for the two models, taking into account both the complexity of the models in the form of the number of its parameters and the fit they achieve with the data in order to obtain a trade-off between the two. The TMCMC algorithm provides the values of the evidence of each model as a by-product of the algorithm. Therefore, by performing model updating on both models they can be easily compared using their evidence values. The log-evidence for the three-parameter flexible-soil model was found to be 2.52, which is slightly less than the evidence value 2.55 of the two-parameter stiff-soil model. Bayesian model selection slightly rewards the stiff-soil model for having one less parameter than the flexible-soil model.

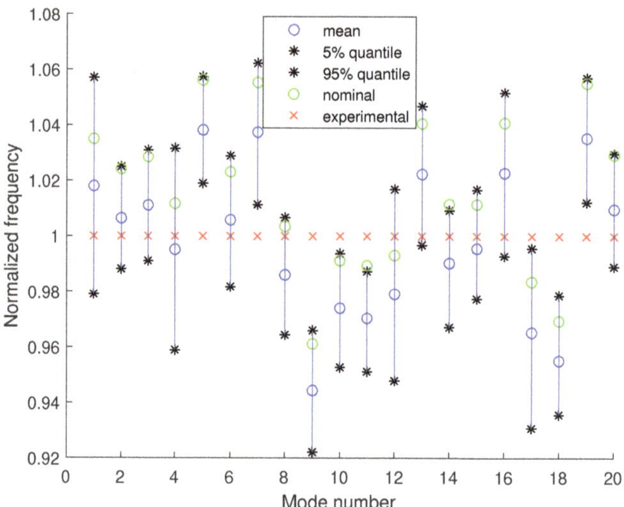

Figure 6. Uncertainty propagation to the first 20 modal frequencies compared with the experimental data and nominal model.

4.2. Model Updating Using Modal Frequencies and Mode Shapes

Next we also include the mode shapes into the dataset used for model updating. Both the vector-based (Section 2.1.1) and the MAC-based (Section 2.1.2) formulations of the likelihood are used to update the deck and pier model parameters of the two-parameter FE model. Regarding the vector-based likelihood formulation, two cases of mode shape component correlation are examined, namely the uncorrelated and exponentially correlated cases. For the exponentially correlated case, two correlation lengths are examined: $\lambda_r = 100$ m and $\lambda_r = 500$ m for all r values.

A crucial aspect of the analysis is to examine the effect of the number of mode shape components (sensors) used in the likelihood function on the model parameter uncertainty and uncertainty in model predictions. In order to study this effect, five different sensor configurations are considered with 8, 14, 26, 52, 105 measured DOF. For each configuration the sensors are selected to be uniformly spread along the bridge deck. In addition, the configuration with a larger number of measured DOF includes the measured DOF contained in configurations with smaller number of DOF. In this way, the information contained in the data of a configuration with a given number of measured DOF, includes the information contained in the data of a configuration with smaller number of DOF. In each sensor configuration case, half of the DOF are transverse (sensors measuring in the transverse direction) and half are vertical. The longitudinal DOF were not included due to their negligible contribution in the identified mode shapes compared to the transverse and vertical components. The transverse and vertical DOF were selected to be in the same point, that is, eight DOF correspond to four different pairs of transverse and vertical DOF in the same point. The case of 52 DOF corresponds to the complete set of measured DOF in the one side of the bridge, while the case of 105 DOFs corresponds to measured DOFs on both sides of the bridge. Due to the type of the vertical and transverse mode shapes, the mode shape components at one side of the bridge provide exactly the same information as the mode shape components at the opposite side of the bridge. So the case of 105 measured DOFs should not be expected to provide additional information as compared to the case of 52 sensors.

Figures 7 and 8 show the posterior parameter uncertainty for the deck and pier parameters of the model as a function of the number of sensors, for each case of likelihood formulation. The posterior uncertainty for each parameter is shown in terms of the 5%, 50% and 95% quantiles of the marginal posterior samples obtained from the TMCMC algorithm for the corresponding parameter.

It should be noted that the vector-based likelihood formulations for the mode shapes (uncorrelated and the two exponentially correlated models with spatial correlation lengths of 100 and 500) result in a steady reduction in the posterior uncertainty of both model parameters (deck and pier parameters), as the number of mode shape components used in the likelihood is increased. This is in agreement with the Bayesian theory of parameter estimation, which suggests that as the number of data points used in the likelihood is increased the posterior uncertainty is reduced. Indeed, in the vector-based formulations, the mode shapes are treated as vectors of size equal to the number of their used components. Therefore, the total number of data points used in the likelihood is increased as we use more of the identified mode shape components

However, as more mode shape components are used, the locations of the sensors become increasingly closer to each other. The shorter characteristic length corresponding to the lowest 10 identified mode shapes is approximately 130 m as one can observe from Figures A1–A4 of Appendix A. As the number of sensors increase to 25 or higher, the shortest distance between sensors becomes a fraction of the characteristic length of the identified mode shapes and so there is redundant information contained in the measured mode shape data. In fact no new information is expected from sensors placed at a distance that is sufficiently smaller than the characteristic length of a mode shape. Especially in the case of 52 and 105 DOFs (which correspond to the entire set of identified mode shape components in one side and both sides of the bridge) we do not expect the inclusion sensor information from the second side to further reduce the posterior uncertainty. This is because the identified mode shape components at the two sides of the bridge are almost identical, and therefore including the second side does not provide any new information about the transverse and vertical mode shapes. The same holds true to a lesser degree for the other cases of DOFs because the sensors are getting closer as we use more of them and contain very similar information. So we would expect the posterior uncertainty to initially reduce as we increase the number of sensors, but only up to a certain point, and then remain practically constant as we include more sensors due to the redundant information provided from the closely spaced sensors or sensors placed at opposite sides of the bridge.

This expected behavior is opposite to what is observed using the vector-based formulations of the likelihood for the mode shapes. Even adding the sensors at the second side of the bridge (which provide identical information with the sensors in one side) seems to further reduce the posterior parameter uncertainty for both the deck and pier parameters. Correlated prediction error models have been suggested to alleviate this situation, and have been successful in some cases, but these correlated prediction error models are very difficult to postulate correctly in practice and could otherwise lead to erroneous results [41].

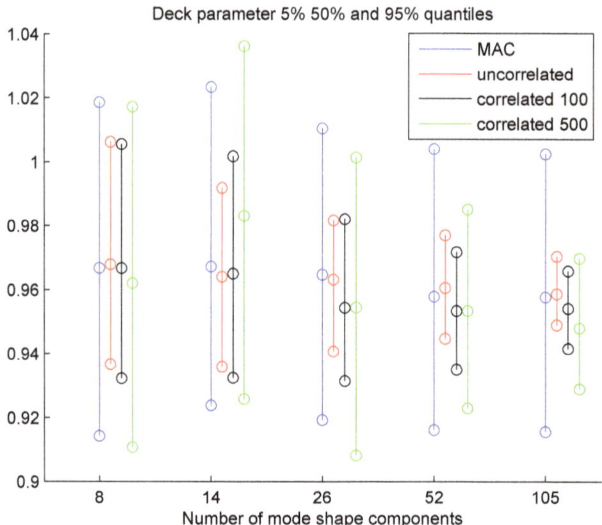

Figure 7. Deck parameter posterior uncertainty versus number of sensors. The MAC case corresponds to the MAC-based likelihood formulation, while "uncorrelated", "correlated 100" and "correlated 200" cases correspond to vector-based likelihood formulations with exponential spatially correlated prediction error given by Equation (8) with $\lambda_r = 0, 100$ and 200 respectively.

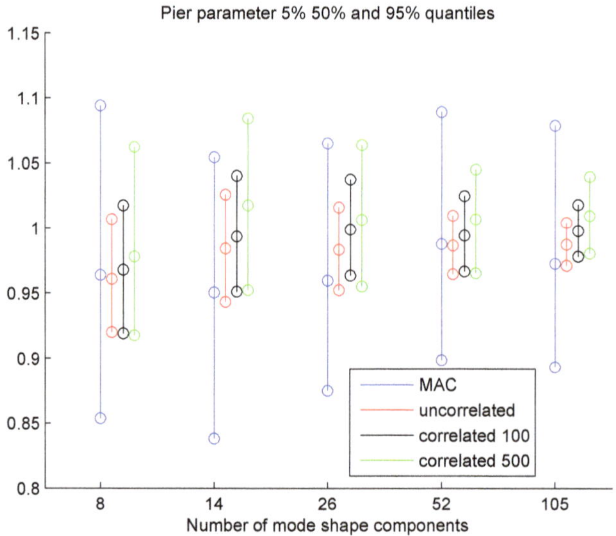

Figure 8. Pier parameter posterior uncertainty versus number of sensors. Description of curves as in Figure 7.

A totally different behavior is observed under the MAC-based likelihood formulation for the mode shapes. The posterior uncertainty does decrease at first, but then it stabilizes and is practically unaffected by the inclusion of more sensors after some point. Specifically, we see that when the number of sensors increase from 8 to 14 and 26, there is a reduction in the uncertainty, but after that point the uncertainty gets stabilized and is not affected by doubling the number of sensors to 52 and eventually to 105. This happens because in the MAC-based formulation the mode shapes are not utilized as vectors, but as scalar MAC values, reducing the effective number of independent data points for each mode shape to one, instead of as many as the number of mode shape components. It is also important to note that the overall parameter uncertainty is much larger compared to the vector-based formulations indicating that no significant information gain occurs by further increasing the number of sensors.

A quantitative assessment is given in Table 4 which shows the 5–95% credible interval for the posterior PDF of the deck and pier model parameters for different number of sensors under the vector-based (uncorrelated) and MAC-based likelihood formulations. It can be seen that the vector-based (uncorrelated) formulation keeps reducing the posterior uncertainty of the model parameters as the number of sensors increase, whereas the uncertainty is maintained for the MAC-based likelihood formulation for 26, 52 and 105 sensors.

Table 4. Deck and pier posterior uncertainty (90% credible interval) versus the number of sensors for the vector-based (uncorrelated) and MAC-based formulations of likelihood.

NDOF	Uncorrelated		MAC	
	Deck	Pier	Deck	Pier
8	0.070	0.087	0.104	0.241
14	0.056	0.083	0.100	0.216
26	0.041	0.064	0.091	0.190
52	0.032	0.045	0.088	0.191
105	0.021	0.033	0.087	0.186

Uncertainty Propagation to Modal Frequencies and MAC Values

The uncertainty in the model parameter values is propagated to modal frequencies in Figure 9 and MAC values in Figure 10. Results are presented for the vector-based (uncorrelated) and MAC-based formulations using 105 DOFs for the mode shapes (sensors on both sides of the bridge). The modal frequency predictions in Figure 9 are normalized with respect to the experimental modal frequencies.

The larger posterior parameter uncertainties obtained with the MAC-based likelihood formulation result in larger uncertainties in the predicted modal frequencies compared to the uncertainties predicted by the vector-based (uncorrelated) likelihood formulation. The experimental frequencies are included within the 5–95% credible intervals predicted by the MAC-based likelihood formulation for 10 out of the 15 modes (the black horizontal line crosses the 5–95% interval except for 5 modes), while the modal frequency predictions obtained from the uncorrelated model do not include the experimental frequencies within the 5–95% credible intervals, except from only three modes (4-th, 6-th and 15-th modes). Therefore, the predictions obtained from the vector-based likelihood formulation have more error associated with them compared with those obtained from the MAC formulation, when checking against the experimentally identified modal frequencies. Thus, the MAC-based likelihood formulation has better predictive capabilities than the vector-based likelihood formulation, in the sense that the predicted uncertainty bounds either fully contain or are closer to the experimental modal frequencies.

The predicted MAC values presented in Figure 10 have also larger uncertainties as expected under the MAC formulation, but the difference is not as obvious as in the modal frequencies (except for mode shapes 11 and 12 which have large uncertainties in their MAC value). The MAC values are well above 0.95 (with the exception of mode shape 9 which has a MAC value of 0.88), indicating a very close match between the experimental and model predicted mode shapes. Note that the MAC values obtained from the nominal model (green circles in Figure 10) are contained within the credible intervals of the vector-based likelihood formulation. This indicates that the mode shapes are highly insensitive to changes in the values of the two model parameters. However, it was demonstrated that inclusion of the mode shapes in the likelihood function does play an important role in the resulting posterior uncertainty of the model parameters, thereby affecting the uncertainty in the predicted modal frequencies.

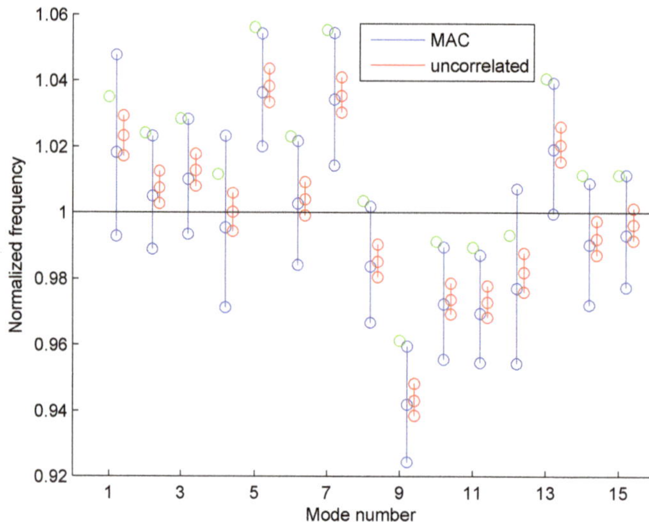

Figure 9. Posterior robust predictions for the modal frequencies normalized by the experimental modal frequencies under the vector-based (uncorrelated) and MAC-based formulations for the likelihood. Vertical lines (with three circles) for each mode number show the 5%, 50% and 95% quantiles. The isolated green circles show predictions from the nominal FE model.

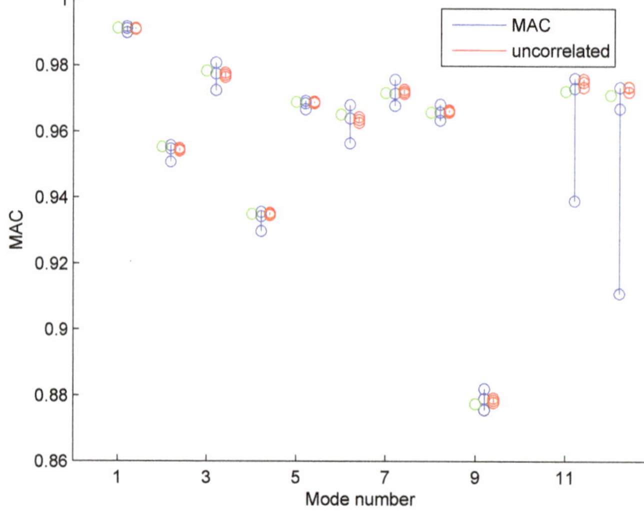

Figure 10. Posterior robust predictions for the mode shapes compared with the experimental mode shapes using the MAC value, under the vector-based (uncorrelated) and MAC-based formulations for the likelihood. Vertical lines (with three circles) for each mode number show the 5%, 50% and 95% quantiles. The isolated green circles show predictions from the nominal FE model.

Based on the uncertainty results for the model parameters presented in Figures 7 and 8, it is expected that the uncertainty in the modal frequencies and mode shapes presented in Figures 9 and 10 will be unaffected for the MAC-based likelihood formulation when the number of sensors is reduced from 102 to 54 or 27. However, for the vector-based likelihood formulation the uncertainty in the modal frequencies and MAC values is expected to increase due to the increase in the parameter uncertainties in Figures 7 and 8 when the number of sensors is reduced from 102 to 54 or 27.

5. Conclusions

A Bayesian framework was presented for FE model updating of structures using experimentally identified modal frequencies and mode shapes. A novel way for including the mode shapes into the likelihood formulation was proposed by assigning a probability model to the MAC values between the experimentally identified and model predicted mode shapes, summarizing the information in the mode shapes in scalar features instead of vectors as it is conventionally done in existing formulations. The MAC-based likelihood formulation provides uncertainty bounds of the model parameters which are consistent with expectations as the number of sensors increases, while the vector-based likelihood formulation fail to properly account for the redundant information contained in the mode shape components, especially for relatively closely spaced sensors. The merits of the new likelihood formulation in relation to existing formulations were explored by updating the FE model of the Metsovo bridge. A high fidelity FE element model of hundreds of thousand of DOF was developed to accurately model the dynamic behavior of the bridge. TMCMC was used to perform the model updating, while model reduction techniques were effectively employed to drastically reduce the computational effort to manageable levels.

It was demonstrated that the model-updating results obtained from the MAC-based likelihood formulation differ significantly from the ones obtained by classical vector-based likelihood formulations. Specifically, the posterior parameter uncertainty was found to be stabilized as the number of sensors in the mode shapes are increased or the distance between sensors is relatively less than the characteristic lengths of the identified mode shapes, or as extra mode shape components (at the opposite side of the bridge), containing redundant information, are added. In contrast, the uncertainty in the model parameters for the classical vector-based likelihood formulation is decreasing as the number of sensors increases, which is counter-intuitive since it does not take into account the redundant information contained in measurements. This decrease in uncertainty is observed for spatially uncorrelated and exponentially correlated prediction error models considered in this study. Propagating the uncertainty in modal frequencies and MAC values, it is demonstrated that the MAC-based likelihood formulation provides wider uncertainty bounds that contain the experimental data. In contrast, the uncertainty bound predicted by the vector-based likelihood formulation fail to fit the experimental data since there is a significant distance of the experimental data from the predicted uncertainty bounds.

Author Contributions: Conceptualization, C.A. and C.P.; methodology, C.A., P.T., P.P. and C.P.; software, C.A.; validation, C.A.; formal Analysis, C.A.; investigation, C.A. and P.P.; resources, C.A. and P.P.; writing—original draft preparation, C.A.; writing—review and editing, C.P., P.T. and P.P.; visualization, C.A.; supervision, C.P. All authors have read and agreed to the published version of the manuscript.

Acknowledgments: We thank Ivan Siu Kui Au for providing a copy of the Bayesian Operational Modal Analysis software used in this study to estimate the modal properties of the bridge.

Conflicts of Interest: The authors declare no conflict of interest.

Appendix A. Supplementary Data

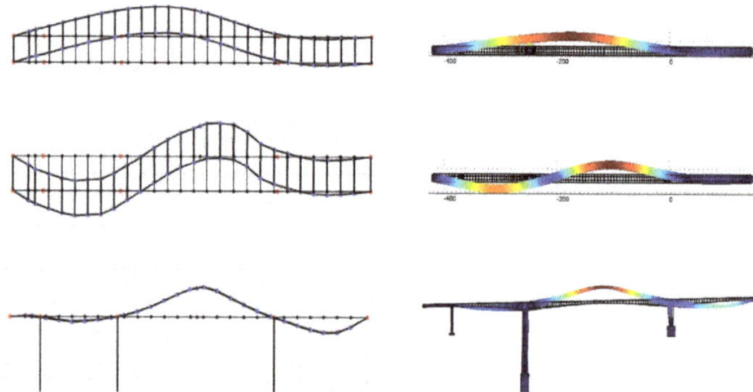

Figure A1. Comparison between the experimentally identified (left column) and nominal FE model predicted (right column) mode shapes of the Metsovo bridge. Modes 1–3.

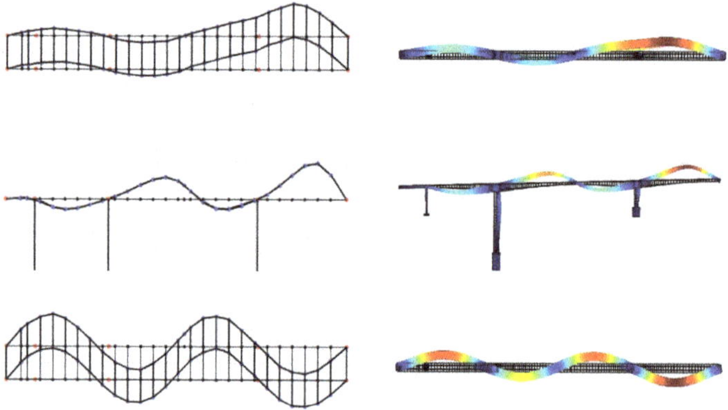

Figure A2. Comparison between the experimentally identified (left column) and nominal FE model predicted (right column) mode shapes of the Metsovo bridge. Modes 4–6.

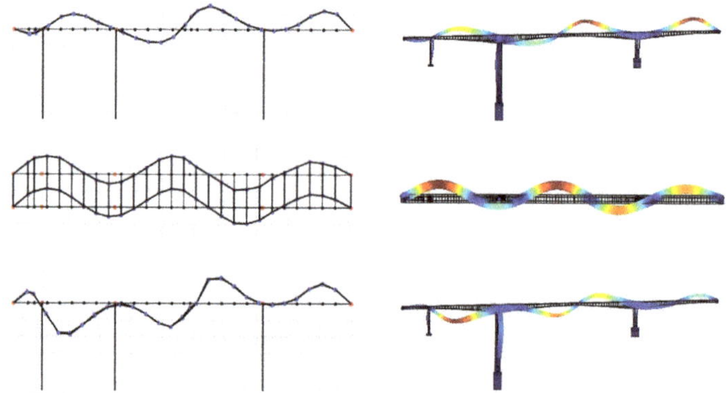

Figure A3. Comparison between the experimentally identified (left column) and nominal FE model predicted (right column) mode shapes of the Metsovo bridge. Modes 7–9.

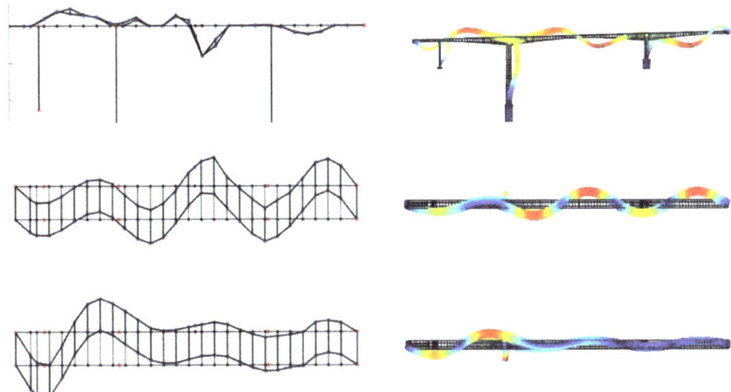

Figure A4. Comparison between the experimentally identified (left column) and nominal FE model predicted (right column) mode shapes of the Metsovo bridge. Modes 10–12.

References

1. Friswell, M.I.; Mottershead, J.E. *Finite Element Model Updating in Structural Dynamics*; Springer Science+Business Media Dordrecht: Berlin, Germany, 1995. [CrossRef]
2. Sehgal, S.; Kumar, H. Structural Dynamic Model Updating Techniques: A State of the Art Review. *Arch. Comput. Methods Eng.* **2015**. [CrossRef]
3. Vanik, M.W.; Beck, J.L. A Bayesian probabilistic approach to structural health monitoring. *Struct. Health Monit.* **1998**, *3243*, 140–151. [CrossRef]
4. Vanik, M.W.; Beck, J.L.; Au, S.K. Bayesian probabilistic approach to structural health monitoring. *ASCE J. Eng. Mech.* **2000**, *126*, 738–745. [CrossRef]
5. Yuen, K.V. *Bayesian Methods for Structural Dynamics and Civil Engineering*; John Wiley & Sons: Singapore, 2010. [CrossRef]
6. Yuen, K.V.; Beck, J.L.; Katafygiotis, L.S. Efficient model updating and health monitoring methodology using incomplete modal data without mode matching. *Struct. Control Health Monit.* **2006**, *13*, 91–107. [CrossRef]
7. Mustafa, S.; Matsumoto, Y. Bayesian model updating and its limitations for detecting local damage of an existing truss bridge. *J. Bridge Eng.* **2017**, *22*, 04017019. [CrossRef]
8. Yuen, K.V.; Beck, J.L.; Katafygiotis, L.S. Probabilistic approach for modal identification using non-stationary noisy response measurements only. *Earthq. Eng. Struct. Dyn.* **2002**, *31*, 1007–1023. [CrossRef]
9. Au, S.K. Fast Bayesian ambient modal identification in the frequency domain, Part I: Posterior most probable value. *Mech. Syst. Signal Process.* **2012**, *26*, 60–75. [CrossRef]
10. Au, S.K. Fast Bayesian ambient modal identification in the frequency domain, Part II: Posterior uncertainty. *Mech. Syst. Signal Process.* **2012**, *26*, 76–90. [CrossRef]
11. Au, S.K.; Zhang, F.L.; To, P. Field observations on modal properties of two tall buildings under strong wind. *J. Wind Eng. Ind. Aerodyn.* **2012**, *101*, 12–23. [CrossRef]
12. Au, S.K.; Zhang, F.L. Ambient modal identification of a primary–secondary structure by Fast Bayesian FFT method. *Mech. Syst. Signal Process.* **2012**, *28*, 280–296. [CrossRef]
13. Au, S.K.; Zhang, F.L.; Ni, Y.C. Bayesian operational modal analysis: Theory, computation, practice. *Comput. Struct.* **2013**, *126*, 3–14. [CrossRef]
14. Zhang, F.L.; Au, S.K. Fundamental two-stage formulation for Bayesian system identification, Part II: Application to ambient vibration data. *Mech. Syst. Signal Process.* **2016**, *66*, 43–61. [CrossRef]
15. Zhang, F.L.; Ni, Y.C.; Au, S.K.; Lam, H.F. Fast Bayesian approach for modal identification using free vibration data, Part I–Most probable value. *Mech. Syst. Signal Process.* **2016**, *70*, 209–220. [CrossRef]
16. Ni, Y.C.; Zhang, F.L.; Lam, H.F.; Au, S.K. Fast Bayesian approach for modal identification using free vibration data, Part II—Posterior uncertainty and application. *Mech. Syst. Signal Process.* **2016**, *70*, 221–244. [CrossRef]
17. Kuok, S.; Yuen, K. Investigation of modal identification and modal identifiability of a cable-stayed bridge with Bayesian framework. *Smart Struct. Syst.* **2016**, *17*, 445–470. [CrossRef]

18. Ni, Y.; Lu, X.; Lu, W. Operational modal analysis of a high-rise multi-function building with dampers by a Bayesian approach. *Mech. Syst. Signal Process.* **2017**, *86*, 286–307. [CrossRef]
19. Beck, J.L.; Au, S.K. Bayesian updating of structural models and reliability using Markov chain Monte Carlo simulation. *J. Eng. Mech. ASCE* **2002**, *128*, 380–391. [CrossRef]
20. Beck, J.L.; Katafygiotis, L.S. Updating models and their uncertainties. I: Bayesian statistical framework. *J. Eng. Mech.* **1998**, *124*, 463–467. [CrossRef]
21. Katafygiotis, L.S.; Beck, J.L. Updating models and their uncertainties. II: Model identifiability. *J. Eng. Mech.* **1998**, *124*, 463–467. [CrossRef]
22. Katafygiotis, L.S.; Papadimitriou, C.; Lam, H.F. A probabilistic approach to structural model updating. *Soil Dyn. Earthq. Eng.* **1998**, *17*, 495–507. [CrossRef]
23. Beck, J.L. Bayesian system identification based on probability logic. *Struct. Control Health Monit.* **2010**, *17*, 825–847. [CrossRef]
24. Yuen, K.V.; Beck, J.L.; Katafygiotis, L.S. Unified probabilistic approach for model updating and damage detection. *J. Appl. Mech. Trans. ASME* **2006**, *73*, 555–564. [CrossRef]
25. Cheung, S.H.; Beck, J.L. Bayesian Model Updating Using Hybrid Monte Carlo Simulation with Application to Structural Dynamic Models with Many Uncertain Parameters. *J. Eng. Mech.* **2009**, *135*, 243–255. [CrossRef]
26. Muto, M.; Beck, J.L. Bayesian updating and model class selection for hysteretic structural models using stochastic simulation. *J. Vib. Control* **2008**, *14*, 7–34. [CrossRef]
27. Goller, B.; Schueller, G.I. Investigation of model uncertainties in Bayesian structural model updating. *J. Sound Vib.* **2011**, *330*, 6122–6136. [CrossRef] [PubMed]
28. Yuen, K.V. Updating large models for mechanical systems using incomplete modal measurement. *Mech. Syst. Signal Process.* **2012**, *28*, 297–308. [CrossRef]
29. Lam, H.F.; Yang, J.; Au, S.K. Bayesian model updating of a coupled-slab system using field test data utilizing an enhanced Markov chain Monte Carlo simulation algorithm. *Eng. Struct.* **2015**, *102*, 144–155. [CrossRef]
30. Yan, W.J.; Katafygiotis, L.S. A novel Bayesian approach for structural model updating utilizing statistical modal information from multiple setups. *Struct. Saf.* **2015**, *52*, 260–271. [CrossRef]
31. Asadollahi, P.; Huang, Y.; Li, J. Bayesian finite element model updating and assessment of cable-stayed bridges using wireless sensor data. *Sensors* **2018**, *18*, 3057. [CrossRef]
32. Au, S.K. Assembling mode shapes by least squares. *Mech. Syst. Signal Process.* **2011**, *25*, 163–179. [CrossRef]
33. Ching, J.; Chen, Y.C. Transitional Markov Chain Monte Carlo method for Bayesian model updating, model class selection and model averaging. *J. Eng. Mech.* **2007**, *133*, 816–832. [CrossRef]
34. Papadimitriou, C.; Papadioti, D.C. Component mode synthesis techniques for finite element model updating. *Comput. Struct.* **2013**, *126*, 15–28. [CrossRef]
35. Jensen, H.A.; Millas, E.; Kusanovic, D.; Papadimitriou, C. Model-Reduction Techniques for Bayesian Finite Element Model Updating Using Dynamic Response Data. *Comput. Struct.* **2014**, *279*, 301–324. [CrossRef]
36. Angelikopoulos, P.; Papadimitriou, C.; Koumoutsakos, P. Bayesian uncertainty quantification and propagation in molecular dynamics simulations: A high performance computing framework. *J. Chem. Phys.* **2012**, *137*, 103–144. [CrossRef]
37. Hadjidoukas, P.; Angelikopoulos, P.; Papadimitriou, C.; Koumoutsakos, P. Π4U: A high performance computing framework for Bayesian uncertainty quantification of complex models. *J. Comput. Phys.* **2015**, *284*, 1–21. [CrossRef]
38. Beck, J.L.; Yuen, K.V. Model selection using response measurements: Bayesian probabilistic approach. *J. Eng. Mech.* **2004**, *130*, 192–203. [CrossRef]
39. Behmanesha, I.; Yousefianmoghadam, S.; Nozaric, A.; Moaveni, B.; Stavridis, A. Uncertainty quantification and propagation in dynamic modelsusing ambient vibration measurements, application to a10-story building. *Mech. Syst. Signal Process.* **2018**, *107*, 502–514. [CrossRef]
40. Simoen, E.; De Roeck, G.; Lombaert, G. Dealing with uncertainty in model updating for damageassessment: A review. *Mech. Syst. Signal Process.* **2015**, *56–57*, 123–149. [CrossRef]
41. Simoen, E.; Papadimitriou, C.; Lombaert, G. On prediction error correlation in Bayesian model updating. *J. Sound Vib.* **2013**, *332*, 4136–4152. [CrossRef]

42. Simoen, E.; Papadimitriou, C.; De Roeck, G.; Lombaert, G. Influence of the prediction error correlation model on Bayesian FE model updating results. In Proceedings of the Life-Cycle and Sustainability of Civil Infrastructure Systems—Proceedings of the 3rd International Symposium on Life-Cycle Civil Engineering (IALCCE 2012), Vienna, Austria, 3–6 October 2012.
43. Papadimitriou, C.; Lombaert, G. The effect of prediction error correlation on optimal sensor placement in structural dynamics. *Mech. Syst. Signal Process.* **2012**, *28*, 105–127. [CrossRef]
44. Papadimitriou, C.; Ntotsios, E.; Giagopoulos, D.; Natsiavas, S. Variability of Updated Finite Element Models and their Predictions Consistent with Vibration Measurements. *Struct. Control Health Monit.* **2012**, *19*, 630–654. [CrossRef]
45. Wu, S.; Angelikopoulos, P.; Papadimitriou, C.; Koumoutsakos, P. Bayesian annealed sequential importance sampling (BASIS): An unbiased version of transitional Markov chain Monte Carlo. *ASCE ASME J. Risk Uncertain. Eng. Syst. Part B Mech. Eng.* **2018**, *4*, 011008. [CrossRef]
46. Craig, R., Jr.; Bampton, M. Coupling of substructures for dynamic analysis. *AIAA J.* **1965**, *6*, 678–685.
47. Jensen, H.; Muñoz, A.; Papadimitriou, C. An enhanced substructure coupling technique for dynamic re-analyses: Application to simulation-based problems. *Comput. Methods Appl. Mech. Eng.* **2016**, *307*, 215–234. [CrossRef]
48. Jensen, H.; Papadimitriou, C. *Sub-Structure Coupling for Dynamic Analysis: Application to Complex Simulation-Based Problems Involving Uncertainty*; Lecture Notes in Applied and Computational Mechanics 89; Springer Nature: Cham, Switzerland, 2019.
49. Castanier, M.; Tan, Y.C.; Pierre, C. Characteristic constraint modes for component mode synthesis. *AIAA J.* **2001**, *39*, 1182–1187. [CrossRef]

Journal of
Sensor and
Actuator Networks

Article

Data-Interpretation Methodologies for Practical Asset-Management

Sai G. S. Pai [1,2,*], Yves Reuland [1] and Ian F. C. Smith [1,2]

[1] Applied Computing and Mechanics Laboratory, School of Architecture, Civil and Environmental Engineering, Swiss Federal Institute of Technology, 1015 Lausanne, Switzerland; yves.reuland@epfl.ch (Y.R.); ian.smith@epfl.ch (I.F.C.S.)
[2] ETH Zurich, Future Cities Laboratory, Singapore-ETH Centre, 1 CREATE Way, CREATE Tower, Singapore 138602, Singapore
* Correspondence: sai.pai@epfl.ch; Tel.: +41-21-69-32498

Received: 23 April 2019; Accepted: 11 June 2019; Published: 22 June 2019

Abstract: Monitoring and interpreting structural response using structural-identification methodologies improves understanding of civil-infrastructure behavior. New sensing devices and inexpensive computation has made model-based data interpretation feasible in engineering practice. Many data-interpretation methodologies, such as Bayesian model updating and residual minimization, involve strong assumptions regarding uncertainty conditions. While much research has been conducted on the scientific development of these methodologies and some research has evaluated the applicability of underlying assumptions, little research is available on the suitability of these methodologies to satisfy practical engineering challenges. For use in practice, data-interpretation methodologies need to be able, for example, to respond to changes in a transparent manner and provide accurate model updating at minimal additional cost. This facilitates incremental and iterative increases in understanding of structural behavior as more information becomes available. In this paper, three data-interpretation methodologies, Bayesian model updating, residual minimization and error-domain model falsification, are compared based on their ability to provide robust, accurate, engineer-friendly and computationally inexpensive model updating. Comparisons are made using two full-scale case studies for which multiple scenarios are considered, including incremental acquisition of information through measurements. Evaluation of these scenarios suggests that, compared with other data-interpretation methodologies, error-domain model falsification is able to incorporate, iteratively and transparently, incremental information gain to provide accurate model updating at low additional computational cost.

Keywords: probabilistic data-interpretation; Bayesian model updating; error-domain model falsification; iterative asset-management; practical applicability; computation time

---

1. Introduction

Improving living conditions and a global trend in migration of population from rural to urban centers result in increasing demand for civil infrastructure [1]. However, most present-day infrastructure has been built in the second half of the twentieth century and is close to the end of its designed service life. The deficit between demand and supply has been estimated to be USD 1 trillion in 2014 [2] and is increasing. Replacement of all aging infrastructure is unsustainable. However, civil infrastructure are generally designed using conservative models and, thus, they may possess reserve capacity beyond code requirements [3,4]. For quantification of such reserve capacity, better understanding of structural behavior is needed. Thus, decision making regarding asset-management actions such as repair, retrofit and replacement, is enhanced [5].

Measurements of structural response can be interpreted using physics-based models in order to enhance understanding of structural behavior. Increased availability and reduced cost of sensing techniques [6,7] and computational tools [8] have made model-based data interpretation feasible. However, all models are idealizations of reality [9]. Conservative modeling assumptions lead to large uncertainty, with systematic and correlated errors at measurement locations [10]. Many researchers have studied uncertainties that affect interpretation of civil-infrastructure response [11–13]. Improvement in quantification of uncertainties can help improve accuracy of data interpretation.

Interpretation of measurements using a physics-based model is referred to as structural identification. Due to the presence of uncertainties, structural identification, which is an abductive task, is an ill-posed problem. Methodologies for solving such inverse problems have been studied by many researchers [14–17]. In practical applications, residual minimization (also called model calibration) is the most commonly used model-based measurement-interpretation method. For residual minimization, optimal values of parameters governing model behavior are estimated by minimizing the residual between model predictions and measurements [18]. Although popular among practicing engineers due to its simplicity, residual minimization may provide inaccurate results [19]. An assumption made by residual-minimization methods is that the difference between model predictions and measurements is governed only by the choice of parameters [20]. This implies that systematic bias between the approximate model and measurements is not taken into account during parameter estimation. In other words, the difference between model predictions and measurements is assumed to be distributed as zero-mean uncertainty forms [10,21–23].

Another methodology that has gathered much interest from the data-interpretation community is Bayesian model updating (BMU). Traditionally, BMU employs an independent zero-mean Gaussian likelihood function [24]. Model parameters, considered as probabilistic distributions are updated using this likelihood function. Model-parameter combinations that provide predictions whose error with measurements are low are attributed higher likelihood. Many developments over the traditional implementation of BMU have been made to account for the presence of model bias [25–27]. However, mis-evaluation of systematic bias and correlations leads to inaccurate estimation of model parameters [28–32].

Goulet and Smith [29] presented a multi-model data-interpretation methodology called error-domain model falsification (EDMF). In EDMF, model-parameter instances are falsified when their predictions are not compatible with measurements. Compatibility is determined based on falsification thresholds that are computed based on the uncertainties affecting identification of parameter values. Estimation of uncertainties involves information available from tests, guidelines and engineering heuristics. EDMF has been shown to provide more accurate identification and predictions compared with traditional BMU and residual minimization [29–32].

Data interpretation for asset management is an iterative task, with re-evaluations required as new information becomes available. New information can be: new measurements; change in uncertainty conditions; and new diagnostic information. Pasquier and Smith [33] presented an iterative sequence-free data-interpretation framework using EDMF and highlighted the iterative nature of data interpretation. A similar framework for post-earthquake assessment using EDMF was presented by Reuland et al. [34]. Zhang et al. [35] developed a data-interpretation tool that employs BMU with the goal of assisting asset managers. Except from these studies, most present day research in structural identification has focused on damage detection within a sequential framework [36]. In addition, no research is available that evaluates the usefulness of data-interpretation methodologies such as BMU and residual minimization within iterative frameworks to assist asset managers faced with real-world challenges. Model-based data-interpretation has the potential to enhance asset management strategies. These strategies have been developed by many researchers using, for example, multi-criteria decision making [37,38] and asset-performance metrics [39,40].

Application of model-based data-interpretation methodologies for full-scale structures presents challenges such as unidentifiability [41] and, above all, constraints on computational cost [42]. Structural identification of full-scale structures, unlike laboratory experiments, are affected by environmental conditions such as wind [43] and temperature [44]. Moreover, numerical models of full-scale structures are computationally expensive. Many researchers have suggested using efficient sampling methods to alleviate constraints on computational cost. Residual minimization has been implemented with optimization algorithms such as genetic algorithms [45], artificial bee colony optimization [46,47], particle swarm optimization [48] and ant colony optimization [49] to reduce computational cost. BMU has been implemented using Markov-Chain Monte Carlo (MCMC) sampling [50], transitional MCMC sampling [51], and evolutionary MCMC sampling [52]. EDMF has traditionally been implemented using grid sampling, which is computationally expensive [53]. Adaptive-sampling strategies such as radial-basis functions [54] and probabilistic global search optimization [55] have been implemented to improve sampling efficiency for EDMF. While use of these search methods decreases computational cost, their efficiency for use in an iterative framework for data-interpretation has not been studied.

In this paper, several methodologies are compared based on their ability to efficiently incorporate new information and changing uncertainty definitions to provide accurate structural identification. Comparisons have been made using two full-scale bridge case studies to evaluate applicability of these data-interpretation methodologies outside of well-controlled laboratory environments.

2. Model-Based Data-Interpretation for Asset Management

Model-based data interpretation of civil infrastructure is difficult due to many scientific and practical challenges. To allow asset managers to exploit potential reserve capacity safely, accurate interpretation of measurement data is necessary. In addition, civil infrastructure (such as bridges and tunnels) form critical components in transportation networks. Their failure can cause loss of life and cascading disruptions to economies due to loss of connectivity. Thus, in addition to accuracy, ease of interpretation of data-interpretation results is imperative to asset managers.

In Figure 1, typical steps involved in using model-based data-interpretation is presented. The sequence shown in the flowchart is general and a sequence-free framework for more specific asset management tasks was presented by Pasquier and Smith [33]. In this paper, the steps involving model-based data-interpretation and validation are discussed.

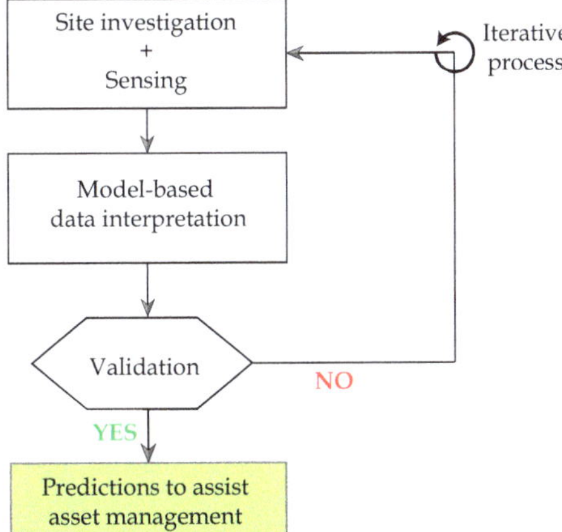

Figure 1. Flowchart detailing typical steps involved in use of model-based data-interpretation methodologies for asset management.

In Figure 1, site investigations help collect data useful for modeling and understanding sources of uncertainty. The task of sensing involves collecting data pertaining to structural response during either a load test or in-service conditions.

The task of model-based data-interpretation in Figure 1 consists of quantification of uncertainties, determination of a model class for identification and use of measurement data to update the values of parameters defined by the model class. Interpretation of measurement data may be carried out using various methodologies and some of these are described in Section 2.1.

A key task that is generally absent in most applications of model-based data interpretation is validation before making predictions. The validation task, subsequent to data interpretation (see Figure 1), has been recommended to be carried out using a leave-one-out cross-validation strategy in this paper. This is explained in Section 2.2.

Data interpretation, as shown in Figure 1, is iterative, especially as new information becomes available over the service life of the structure. New information takes the form of new measurements, improved understanding of uncertainties, improved understanding leading to a new model class, etc. Lack of validation of data-interpretation results may also necessitate iterations. Therefore, methodologies for data interpretation should be amenable to new information and flexible to changes.

In Section 2.1, data-interpretation methodologies that are available in the literature are described. These methodologies make implicit assumptions regarding estimation and quantification of uncertainties affecting structural identification. Methodologies also differ in the sampling strategies used to obtain appropriate solution(s). Accuracy of these methodologies in interpreting measurement data is dependant on the validity of assumptions. A leave-one-out cross-validation method to assess accuracy and precision of data-interpretation is explained in Section 2.2. In this paper, a comparison of data-interpretation methodologies based on iterative applications is presented. The objective of these comparisons and validation checks are to help engineers select a suitable methodology to interpret measurements using physics-based models.

2.1. *Background of Data-Interpretation Methods*

In this paper, four data-interpretation methodologies are compared with respect to their ability to provide accurate identification and incorporate new information in an iterative manner. In the following, the four methodologies are briefly introduced and their inherent assumptions are discussed.

2.1.1. Residual Minimization

In residual minimization, a structural model is calibrated by determining model-parameter values that minimize the error between model predictions and measurements. A typical objective function for residual minimization is shown in Equation (1).

$$\hat{\theta} = \underset{\theta}{\operatorname{argmin}} \sum_{i=1}^{n_y} \left(\frac{g_i(\theta) - \hat{y}_i}{\hat{y}_i} \right)^2 \tag{1}$$

In Equation (1), $\hat{\theta}$ is the optimum model-parameter set obtained using measurements and $g_i(\theta) - \hat{y}_i$ is the residual obtained between the model response, $g_i(\theta)$, and measurement, $\hat{y}_i$, at measurement location i.

Predictions with models updated using residual minimization are limited to the domain of data used for calibration [56]. Therefore, calibrated model-parameter values may only be suitable for predictions that involve interpolation [56] and not for extrapolation (predictions outside the domain of data used for calibration) [19,20].

2.1.2. Traditional Bayesian Model Updating

Bayesian model updating (BMU) is a popular probabilistic data-interpretation methodology [24,57,58] based on Bayes' theorem. In BMU, prior information of model parameters, $P(\theta)$, is conditionally updated using a likelihood function $P(y \mid \theta)$ to obtain a posterior distribution of model parameters, $P(\theta \mid y)$, as shown in Equation (2).

$$P(\theta \mid y) = \frac{P(y \mid \theta) \cdot P(\theta)}{P(y)} \tag{2}$$

In Equation (2), $P(y)$ is the normalization constant. $P(\theta)$ is the prior distribution of model parameters, which indicates prior available knowledge regarding parameter values. The likelihood function, $P(y \mid \theta)$, is the probability of observing the measurement data, y, for a specific set of model-parameter values, θ. The most commonly used likelihood function is a L_2-norm-based Gaussian probability-distribution function (PDF), as shown in Equation (3).

$$P(y \mid \theta) \propto constant \cdot \exp\left[-\frac{1}{2}\left(g\left(\theta\right)-y\right)^{\mathrm{T}}\Sigma^{-1}\left(g\left(\theta\right)-y\right)\right] \tag{3}$$

In Equation (3), Σ is a covariance matrix that consists of variances and correlation coefficients of uncertainties related to each measured location. In most applications of BMU, uncertainties at measurement locations are assumed to be independent zero-mean Gaussian distributions [59–66]. In addition, the variance in uncertainty, σ^2, is assumed to be the same for all measurement locations. This leads the covariance matrix to be a diagonal matrix, with all non-zero terms being equal. However, the assumption of a L_2-norm-based Gaussian distribution for uncertainty [67] and uncorrelated error [28] is rarely satisfied and may lead to a biased updated probability distribution [29,30,32].

2.1.3. Error-Domain Model Falsification

Error-domain model falsification (EDMF) is a data-interpretation methodology developed by Goulet and Smith [29]. EDMF is based on the assertion by Popper [68] that models cannot be validated by data; they can only be falsified. Model instances (instances of model-parameter values) are falsified based on information from measurements. Model instances that are not falsified form a candidate set, which is a subset of all possible parameter values based on the prior model parameter PDFs.

Generally, civil infrastructure are designed using conservative and simplified models. As a result, engineering models possess significant model bias from sources such as simplification of loading conditions, geometrical properties, material properties and boundary conditions. Extent of these uncertainties can only be estimated using engineering heuristics and usually takes the form of bounds.

Let $\epsilon_{mod,q}$ be the modeling uncertainty and $\epsilon_{meas,q}$ the measurement uncertainty, both at a measurement location q. Let the structure be represented by a physics-based model, $g(\theta)$. The true response of the structure at a measurement location is given by Equation (4).

$$y_q + \epsilon_{meas,q} = R_q = g_q\left(\theta^*\right) + \epsilon_{mod,q} \tag{4}$$

In Equation (4), $g_q\left(\theta^*\right)$ is the model response at a measurement location q for the real values of the model parameters, θ^*. y_q is the measured response of the structure at measurement location q. Rearranging the terms of Equation (4), a relationship among model response, $g(\theta)$, measurement, y_q, and uncertainties, $\epsilon_{meas,q}$ and $\epsilon_{mod,q}$, at location q is obtained, as shown by Equation (5).

$$g_q\left(\theta^*\right) - y_q = \epsilon_{meas,q} - \epsilon_{mod,q} \tag{5}$$

In Equation (5), the residual between model response, $g(\theta)$, and measurement, y_q, at a sensor location, q, is equal to the combined model and measurement uncertainty. Engineers make design decisions based on predefined target reliability. Using the target reliability, ϕ, for identification, the criteria for falsification, thresholds $T_{high,q}$ and $T_{low,q}$, are computed using Equation (6).

$$\phi^{1/m} = \int_{T_{low,q}}^{T_{high,q}} f_{U_{c,q}}\left(\epsilon_{c,q}\right) d\epsilon_{c,q} \tag{6}$$

In Equation (6), $f_{U_{c,q}}\left(\epsilon_{c,q}\right)$ is the PDF of combined uncertainty at measurement location q and ϕ is the target reliability of identification. Thresholds, $T_{high,q}$ and $T_{low,q}$, correspond to the shortest interval providing a probability equal to target reliability, ϕ. In Equation (6), the term $1/m$ is the Šidák correction [69], which accounts for m independent measurements used in identification of model parameters. In EDMF, compatibility of model predictions with measurements at each sensor location is considered as a hypothesis, which can have false positives and negatives. Inclusion of false positives as a candidate instance decreases precision of model updating, while falsely rejecting a model instance could potentially lead to falsification of true parameter values. Šidák correction controls the error rate such that the possibility of rejecting the true parameter values is lower than $1 - \phi$.

EDMF is traditionally carried out using grid sampling. In grid sampling, samples from prior distribution of model parameters, $\boldsymbol{\theta}$, are drawn. If n_s samples are drawn from the prior distribution of each parameter, then these samples constitute a grid, which is called the initial model set (IMS). For n_s samples drawn from n_p parameters, the total number of model instances in the IMS is $n_s^{n_p}$.

Residual between model responses, $g(\theta)$, and measurements, y, are compared with the thresholds, $T_{low,q}$ and $T_{high,q}$. If the residual between model response and measurements lies within the thresholds for all measurement locations, then the model instance is accepted. This criteria for falsification is shown in Equation (7).

$$T_{low,q} \leq g_q(\theta) - y_q \leq T_{high,q} \qquad q \in \{1...m\} \tag{7}$$

If predictions for a model instance, θ_i, does not satisfy Equation (7) for even one measurement location, then that model instance is falsified. All candidate model instances are considered equally likely and, thus, assigned a uniform probability density. Candidate models are used for making further predictions using the physics-based model with reduced parametric uncertainty [30]. The EDMF methodology has been applied to more than 20 full-scale systems since 1998 [4]. Recent applications include: model identification [70]; leak detection [71,72]; wind simulation [73]; fatigue life evaluation [74–76]; measurement-system design [77–80]; post-earthquake assessment [34,81]; damage localization in tensegrity structures [82]; and occupant localization [83].

EDMF when compared with BMU and residual minimization has been shown to provide accurate identification due to its robustness to correlation assumptions and explicit estimation of model bias based on engineering heuristics [29–32]. Although grid sampling carries some advantages with respect to practical applications and parallel computing, grid sampling remains computationally expensive [53].

2.1.4. Modified Bayesian Model Updating

To alleviate shortcomings of traditional BMU (see Section 2.1.2), a box-car likelihood function is utilized for modified BMU, which is more robust to incomplete knowledge of uncertainties and correlations than standard L_2-norm-based Gaussian likelihood functions. The box-car likelihood function is developed using an L_∞-norm-based Gaussian likelihood function [67], which is defined as shown in Equation (8).

$$L(\boldsymbol{y} \mid \boldsymbol{\theta}) = \begin{cases} 1/2\sigma_\infty & \text{for } \boldsymbol{\mu_y} - \sigma_\infty \leq \boldsymbol{g(\theta)} - \boldsymbol{y} \leq \boldsymbol{\mu_y} + \sigma_\infty, \\ 0 & \text{otherwise.} \end{cases} \tag{8}$$

In Equation (8), parameters of the likelihood function, $\boldsymbol{\mu_y}$ and σ_∞, are determined using Equations (9) and (10).

$$\boldsymbol{\mu_y} = \frac{\boldsymbol{T_{high}} + \boldsymbol{T_{low}}}{2} \tag{9}$$

$$\sigma_\infty = \boldsymbol{T_{high}} - \boldsymbol{\mu_y} \tag{10}$$

In Equations (9) and (10), $\boldsymbol{T_{low}}$ and $\boldsymbol{T_{high}}$ are the thresholds computed for EDMF using Equation (6) for a target reliability of identification ϕ_d. Modified BMU using such a box-car likelihood distribution has been shown to provide results similar to those obtained using EDMF [31,32].

2.2. Practical Challenges Associated with Model-Based Data Interpretation

As stated before, data interpretation is an iterative task that requires exploring results and re-evaluating results in light of new information regarding uncertainties or new measurements. In addition, the task of data-interpretation may have to be repeated when identification results are found to be inaccurate due to a wrong model class. Assessing accuracy is a challenge as knowledge of true parameter values is unavailable. Accuracy can be approximated with cross-validation methods [84]. While enabling accuracy estimation, such validation strategies are limited to the domain of data used for identification. Moreover, when these methods indicate that structural identification is inaccurate, diagnostics are required to re-assess the assumptions made during identification.

Cross validation of structural-identification results can be conducted using several techniques, such as leave-one-out, hold-out and k-fold. Hold-out and k-fold cross-validation require large measurement datasets for identification and validation. In structural identification of civil infrastructure, measurements are typically scarce with few sensors that provide information about structural behavior. Thus, leave-one-out cross-validation is preferred over other strategies for validating model-updating results.

In leave-one-out cross validation, observation from one sensor is omitted (left-out) and structural identification is carried out using all remaining measurements. Updated model-parameter values are then used to predict the model response at the omitted sensor. If the omitted measurement is compatible with updated model predictions, then structural identification is concluded to be accurate for that sensor location. This procedure is repeated by omitting each sensor separately in order to assess accuracy at all measurement locations.

Consider that n_m measurements are acquired during a load test. These measurements are used for updating a physics-based model of the structure, $g(\boldsymbol{\theta})$, which has parameters $\boldsymbol{\theta} = [\theta_1, \theta_2, ..., \theta_{n_p}]$, where n_p is the number of parameters. Prior to model updating, the initial prediction at a sensor location j is given by Equation (11).

$$q_j = g_j\left(\boldsymbol{\theta}\right) + \epsilon_{pred,j} \tag{11}$$

In Equation (11), $g_j\left(\boldsymbol{\theta}\right)$ is the model prediction at sensor j for model parameters $\boldsymbol{\theta}$ and $\epsilon_{pred,j}$ is the model error from sources such as parameters not considered in the parameter vector $\boldsymbol{\theta}$, uncertainty in load and its position. The model-prediction distribution including model error before model updating is q_j.

Let measurement from sensor j be excluded from model updating of parameters $\boldsymbol{\theta}$ to perform leave-one-out cross-validation. The Sidâk correction for determining the threshold bounds, leaving one sensor out, is $\frac{1}{n_m-1}$. Model parameters, $\boldsymbol{\theta}$, are updated to obtain candidate model parameters, $\boldsymbol{\theta}''$. Model updating is performed using the four methodologies described in the previous sections. The following equations apply directly to EDMF and modified BMU. For traditional BMU, they can be calculated based on the 95th-percentile bound of the prediction distribution. Updated parameters, $\boldsymbol{\theta}''$, are provided as input to the physics-based model, $g(\boldsymbol{\theta}'')$, to predict the model response at the omitted sensor, j, as shown in Equation (12).

$$q_j'' = g_j\left(\boldsymbol{\theta}''\right) + \epsilon_{pred,j} \tag{12}$$

In Equation (12), q_j'' is the distribution of updated model predictions at sensor location j. Depending on the uncertainties and relationships between model parameters and response, the prediction distributions obtained using Equations (11) and (12) are irregular. They are assumed to have uniform distributions based on the principle of maximum entropy. When bounds of the updated distribution of model predictions include the measured value, which has been left out for model updating, then identification is considered to be accurate.

Using leave-one-out cross validation, precision of structural identification can be quantified in addition to accuracy. Precision is a measure of variability either in updated model-parameter distributions or model predictions. Using leave-one-out cross-validation precision is estimated using the error between the updated model-prediction distributions and corresponding measurements. The model-prediction distribution at sensor j before and after model updating is given by Equations (11) and (12). The measurement at this sensor location is y_j. The prediction error is the residual between model predictions and measurement. The prediction error distribution at sensor j, before and after model updating are uniform and their ranges are given by Equations (13) and (14).

$$\mathfrak{R}_j = \frac{q_{j,max} - q_{j,min}}{y_j} \tag{13}$$

$$\mathfrak{R}''_j = \frac{q''_{j,max} - q''_{j,min}}{y_j} \tag{14}$$

In Equations (13) and (14), $\mathfrak{R}_j$ and $\mathfrak{R}''_j$ are ranges of prediction-error distributions relative to the measurement at sensor j before and after model updating. For n_m cases of leave-one-out cross-validation, n_m prediction-error ranges before and after model updating are obtained. Considering prediction-error ranges for all cases of leave-one-out cross-validation before and after model updating, the precision index, φ, is defined as shown in Equation (15).

$$\varphi = \frac{(\mu_{\mathfrak{R}} - \mu_{\mathfrak{R}''})}{\mu_{\mathfrak{R}}} \tag{15}$$

In Equation (15), $\mu_{\mathfrak{R}}$ and $\mu_{\mathfrak{R}''}$ are the mean of prediction-error ranges, before and after model updating, over all cases of sensors left out. Precision, φ, represents reduction in prediction error after model updating and ranges from 0 to 1. Precision, φ, equal to zero implies no gain in information from model updating. In such situations, $\mu_{\mathfrak{R}}$ is equal to $\mu_{\mathfrak{R}''}$, implying that on average over all cases of leave-one-out cross-validation, no reduction in prediction uncertainty is obtained. Precision index, φ, equal to one implies perfect model updating wherein updated parameter and consequently the prediction distributions have zero variability.

3. Full-Scale Applications

In this section, with the help of two full-scale case studies, the application of four data-interpretation methodologies is compared. Comparisons are made with respect to their ability to provide accurate identification and transparently and iteratively incorporate new information.

3.1. Ponneri Bridge

Ponneri Bridge, shown in Figure 2b, is a steel railway bridge located close to Chennai, India. It is part of a system of bridges (see Figure 2a) built in 1977 that comprises a railway crossing over the Arani river.

(a) System of bridges over the Arani river

(b) Instrumented bridge (Ponneri)

Figure 2. (**a**) System of railway-bridges over the Arani river in Chennai; and (**b**) instrumented bridge evaluated in this section, which is called Ponneri Bridge.

The behavior of each bridge in this system is independent of the others. One of the bridges in this system, referred to as Ponneri Bridge, is instrumented. This is the first bridge in the system for a train entering from Chennai and heading north. The bridge has a span of 18.3 m and is composed of two steel I-section girders, as shown in Figure 3. The two steel girders are connected through diagonal cross-bracing with a spacing of 1.6 m, which provides the bridge with stiffness in the transversal direction.

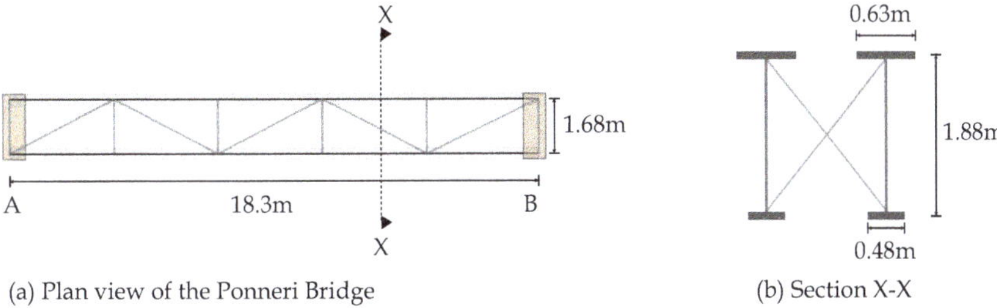

Figure 3. (**a**) Plan view of the bridge; and (**b**) section X-X showing details of the two steel girders.

A finite-element (FE) model of the Ponneri Bridge was developed in Ansys [85]. In the model, the steel girders were modeled using SHELL182 elements. Diagonal bracings (see Figure 3) connecting the steel girders are modeled using BEAM188 elements. The bridge was modeled as simply supported with perfect pin-conditioned support at end B (see Figure 3a) and a partially pin-conditioned support at end A. At end A, the support was modeled to have infinite vertical stiffness with stiffness in longitudinal direction (along the span) parameterized using zero-length spring elements, COMBIN14 (roller support). The bounds of this parameter, $K_{A,z}$, was estimated using the FE model and is reported in Table 1. Other than stiffness of support at end A in longitudinal direction, the Young's modulus of steel, E_s, was parameterized. The bounds for E_s, reported in Table 1, were conservatively estimated based on the probabilistic model for modulus of elasticity of steel provided by Vrouwenvelder [86].

Using the FE model, strain measurements for the bridge were simulated at 16 locations, which are shown in Figure 4. Use of simulated measurements helped compare accuracy of model updating for various scenarios using four data-interpretation methodologies.

Table 1. Prior uncertainty distributions of parameters included in the FE model. The model parameters were assumed to have a uniform distribution (U).

Parameter	Distribution
Young's modulus of elasticity of steel, E_s (GPa)	U(195, 215)
Longitudinal stiffness of support at end A, $k_{A,z}$ (log N/mm)	U(3, 6)

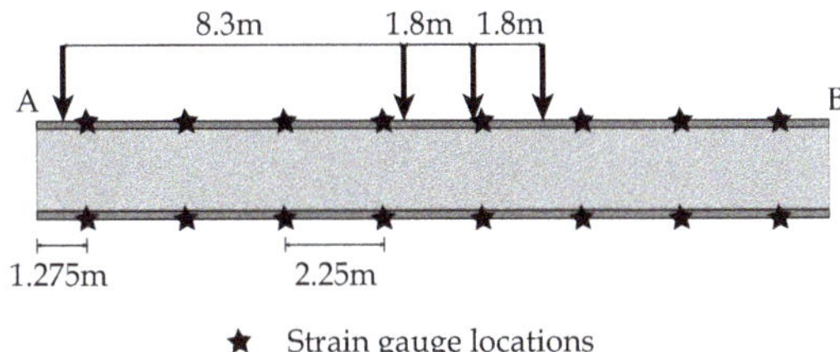

Figure 4. Location of strain gauges for which measurements were simulated using a FE model of the Ponneri Bridge.

When simulating measurements using the FE model, model bias was introduced by assigning partial rotational rigidity to the boundary conditions at supports A and B. Other than the model bias introduced, the values of model parameters, E_s and $k_{A,z}$, assigned to simulate measurements for two conditions of the bridge are shown in Table 2.

Table 2. True values of model parameters E_s and $k_{A,z}$ used to simulate measurements.

Condition	E_s (GPa)	$k_{A,z}$ (log N/mm)
Before retrofit	210	4
After retrofit	210	7

The before-retrofit scenario in Table 2 is the present condition of the bridge. The bridge was then assumed to be retrofitted by replacement of bearing at support A. The new bearing prevents translational movement of the bridge, similar to the bridge bearing at support B. This was represented by increased stiffness of the support A in the longitudinal direction in Table 2. For the two scenarios reported in Table 2, measurements were simulated for a train passing over the bridge. The train was positioned on the bridge such that it produces maximum strain at sensors close to mid-span, as shown in Figure 4.

Using the two scenarios of the Ponneri Bridge (see Table 2), the applicability of four data-interpretation methodologies to provide accurate structural identification was compared. As measurements were simulated, knowledge of the "true" parameter values helps assess identification accuracy in the presence of systematic bias.

Identification of parameters, E_s and $K_{A,z}$, had to be carried out in the presence of uncertainty from multiple sources, such as model bias (due to partial rotational stiffness of end conditions) and sensor noise (added to measurements). Estimations of these uncertainties are shown in Table 3.

The two sources of uncertainties shown in Table 3 were combined using Monte Carlo sampling to determine the updating criteria for traditional BMU, modified BMU and EDMF. For traditional BMU, a zero-mean L_2-norm-based Gaussian likelihood function as defined by Equation (3) was developed. Falsification thresholds for EDMF were calculated using Equation (6). L_∞-norm-based Gaussian likelihood function for modified BMU was developed using Equations (8)–(10). Using the Ponneri case study, the four data-interpretation methodologies were compared with respect to their ability to provide accurate structural identification as well as their amenability to incorporate changes in uncertainty definitions with respect to four identification scenarios, as shown in Table 4.

Table 3. Distribution of uncertainty sources affecting structural identification. Uncertainties were estimated relative (%) to design model predictions.

Scenario	Model Bias	Measurement Uncertainty
1	U(−38, 2)	N(0, 2)
2	U(−40, 8)	N(0, 2)

Table 4. Cases of structural identification considered for comparison of four data-interpretation methodologies.

Scenario	Condition	Description
1	Before retrofit	Without model bias
2	Before retrofit	With model bias
3	After retrofit (replacement of bearing)	Without re-evaluating prior PDFs
4	After retrofit	After re-evaluating prior PDFs

3.1.1. Scenario 1: Structural Identification before Retrofit, Ignoring Model Bias

Structural identification of E_s and $k_{A,z}$ was conducted with measurements recorded before retrofit of the Ponneri Bridge. The first iteration of data interpretation was carried out without taking into account model bias (Scenario 1 in Table 4).

For EDMF, an initial grid of model instances was generated based on the prior distribution of model parameters (see Table 1). A total of 273 model instances (13 $k_{A,z}$ instances multiplied by 21 E_s instances drawn from the prior parameter distributions) were generated and provided as input to the FE model. Model instances whose responses were compatible with measurements at all sensor locations were accepted as part of the CMS. Falsification thresholds obtained by ignoring the model bias falsified all model instances. Thus, the model class is falsified. Falsification of the entire model class indicates mis-evaluation of uncertainties.

Modified BMU was conducted with a boxcar-shaped likelihood function (see Equation (8)). Samples from the joint posterior PDF of model parameters were drawn using MCMC sampling. The starting point was determined using MC sampling. Without taking the model bias into account in development of the likelihood function, after 1000 MC samples, no starting point was obtained, which suggests rejection of the model class by modified BMU, in a similar way to EDMF.

Traditional BMU was conducted with a zero-mean Gaussian likelihood function, without taking into account the model uncertainty. The posterior PDF thus obtained was precise and biased from the true parameter values used to simulate the measurements. Similar to traditional BMU, residual minimization returned parameter values, 215 GPa and 4.5 log N/mm, as optimal. These values were inaccurate and biased from the true parameter values (see Table 2).

3.1.2. Scenario 2: Structural Identification before Retrofit, Considering Model Bias

Based on diagnostics obtained using EDMF and modified BMU, data-interpretation was repeated by including model bias (Scenario 2 in Table 4). EDMF, with re-calculated falsification thresholds and without requiring any new simulations, provided the CMS shown in Figure 5a.

Modified BMU was repeated by taking into account the model bias. Due to limitations on computational cost in using a FE model of a full-scale bridge, only 500 samples were drawn using MCMC sampling. A scatter plot of the samples drawn from the joint posterior PDF of model parameters, E_s and $k_{A,z}$, is shown in Figure 5b.

Traditional BMU was also repeated with a zero-mean Gaussian likelihood function using MCMC sampling (500 samples). A scatter plot of the samples drawn is shown in Figure 5c.

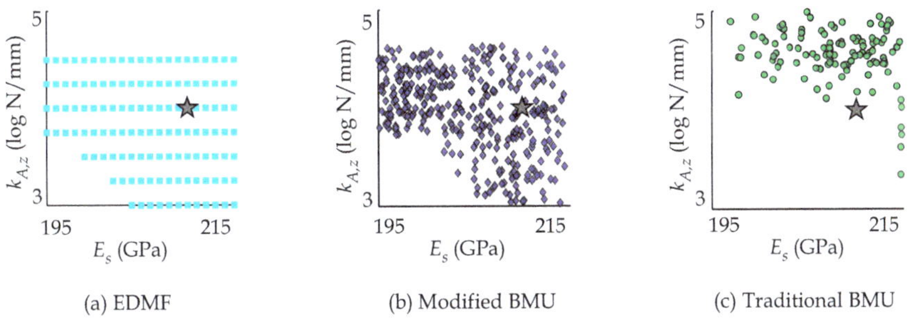

Figure 5. Samples from updated parameter distributions obtained using: (**a**) EDMF; (**b**) modified BMU; and (**c**) traditional BMU before retrofit of the Ponneri Bridge.

In Figure 5, the values of parameters used to simulate measurements are also shown for comparison. For EDMF and modified BMU, the "true" parameter values lie within the updated values of model parameters obtained, as shown in Figure 5a,b. However, for traditional BMU, the updated posterior distribution of model parameters does not include the "true" parameter values (see Figure 5c). Therefore, EDMF and modified BMU provide accurate structural identification for the Ponneri Bridge before retrofit actions, while traditional BMU does not. Since uncertainties were ignored in residual minimization, it was not repeated for changes in estimation of uncertainties.

3.1.3. Scenario 3: Structural Identification after Retrofit, without Re-Evaluating Prior Parameter Distributions

Simulated measurements after retrofit were used for structural identification using the four data-interpretation methodologies (Scenario 3 in Table 4). Replacement of the bearing at support A increased the stiffness of the support in longitudinal direction. The parameter values used to simulate the bridge response (see Table 2) lie outside the bounds of the prior distribution of model parameter $k_{A,z}$ (see Table 1). Therefore, EDMF falsified the entire model class, which implies either uncertainty or other assumptions have to be re-evaluated. Similar to EDMF, modified BMU failed to find a starting point, which suggests that no parameter values from the prior distribution of model parameters have a non-zero likelihood.

Traditional BMU using MCMC sampling found a starting point and identified a posterior with low variability (high precision). As uncertainties affecting structural identification were not changed compared with Scenario 2, this precision was not due to decrease in uncertainty rather mis-evaluation of the prior distributions of model parameters, which have to be re-evaluated.

EDMF, in addition to falsifying the model class, provided diagnostics to re-assess the uncertainty definitions. A comparison of rejected model predictions at all sensor locations with falsification thresholds suggests that the prior distribution of model parameters have to be re-evaluated. Moreover, the error between model response and the thresholds is not the same at all sensor locations, suggesting a systematic source of uncertainty. The only systematic uncertainty source in the model class is $K_{A,z}$ and the prior distribution of this parameter was modified to be uniform with bounds [3, 8] log N/mm.

3.1.4. Scenario 4: Structural Identification after Retrofit, after Re-Evaluating Prior Parameter Distributions

Model parameters were identified with new prior distributions using the four data-interpretation methodologies. The updated parameter distributions obtained using the three probabilistic data-interpretation methodologies are shown in Figure 6 (Scenario 4 in Table 4).

In Figure 6, a scatter plot of samples from the updated parameter distributions is shown. Using post-retrofit measurements of the Ponneri Bridge, all three probabilistic data-interpretation methodologies provided accurate updated parameter distributions. Residual minimization provided 215 GPa and 7 log N/mm as optimal parameter values, which were biased from the "true" parameter values (see Table 2).

To summarize, EDMF and modified BMU provide accurate model updating for all scenarios, before and after retrofit (Cases 1–4 in Table 4). Traditional BMU (considering 95th percentile bounds) provided accurate identification only post-retrofit with appropriate estimation of priors (Case 4 in Table 4). Residual minimization did not provide accurate model updating for any scenario. The various data interpretation scenarios were typical iterations due to unsuccessful validation, as described in Figure 1.

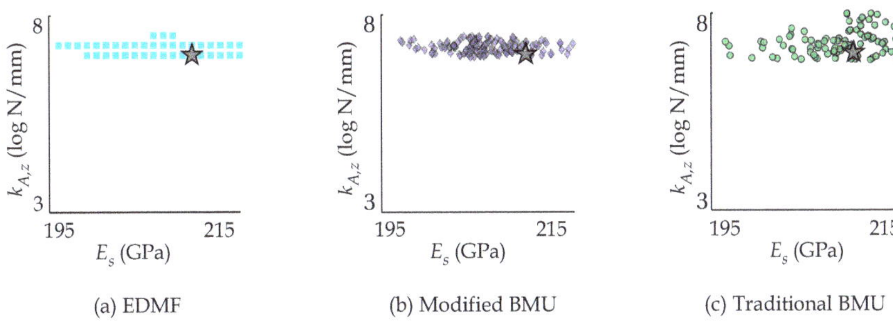

Figure 6. Samples from updated parameter distributions obtained using: (**a**) EDMF; (**b**) modified BMU; and (**c**) traditional BMU after retrofit of the Ponneri Bridge.

3.1.5. Interpretation of Identification Results

Residual minimization is the most commonly used data-interpretation methodology in practice, due to the simplicity of updating criteria (calibration) and ease of result interpretation (single optimal values). However, residual minimization is unable to provide accurate model updating in presence of large modeling uncertainties. Probabilistic data-interpretation methodologies account for the presence of modeling uncertainties, thereby improving accuracy, sometimes at the cost of transparency in understanding model-updating results.

EDMF provides a set of candidate models that are compatible with measurements. Modified BMU provides a bounded joint posterior PDF. Thus, posterior obtained using modified BMU can also be interpreted as updated bounds on model parameters, similar to those obtained using EDMF. Traditional BMU using uniform priors and a L_2-norm-based Gaussian likelihood function provides an informed (not uniform) joint PDF as posterior. The normality of the posterior PDF is dependent on the information gained from measurements. In Figure 7, a comparison of results obtained from model updating using the three probabilistic methodologies for parameter $K_{A,z}$ is shown.

In Figure 7a, updated bounds of the parameter $K_{A,z}$ obtained using EDMF are highlighted. Using these bounds, an engineer is able to interpret the structural behavior for making further predictions to assist asset-management. For example, the updated bounds in Figure 7a indicate that the boundary-condition stiffness (in longitudinal direction) is low, thus the engineer can assume the structure to have a roller support. With this assumption, an engineer is able to make predictions with appropriate additional uncertainty from updated parametric variability. Modified BMU also provides updated bounds of the parameter $K_{A,z}$, as shown in Figure 7b.

Traditional BMU, unlike EDMF and modified BMU, provides an informed posterior PDF, as shown in Figure 7c, where the marginal posterior PDF of model parameter $K_{A,z}$ is highlighted. However, informed PDFs do not directly assist engineers in interpreting results with respect to their physical meaning. Post-processing of results, such as calculation of MAP or 95th-percentile bounds, is necessary for engineers to interpret the physical meaning of interpretation results. With such post-processing, the engineers can determine physical meaning of results obtained from model updating and subsequently use them to make further predictions. In addition, as shown in Figure 7, traditional BMU does not provide accurate identification of the parameter, $K_{A,z}$. An updated understanding of structural behavior, provided as bounds, helps engineers decide upon model-parameter values to be chosen for further predictions in a transparent manner and also reduce computational cost.

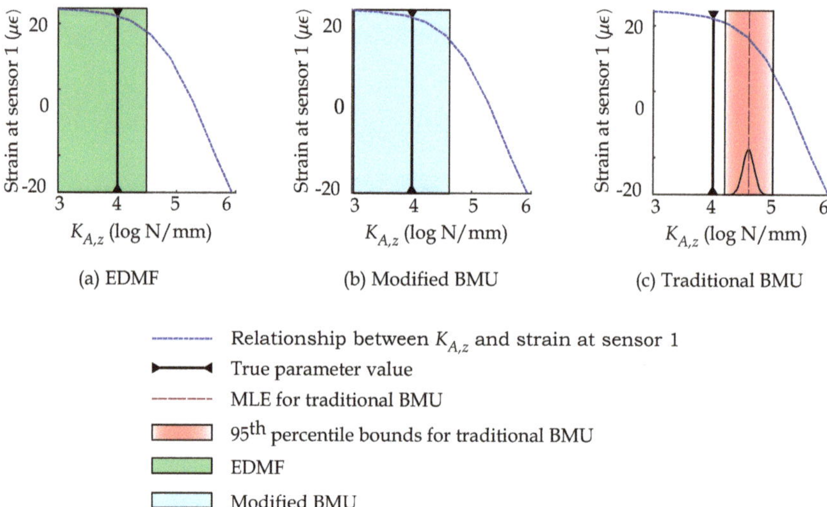

Figure 7. Updated knowledge of parameter $K_{A,z}$ obtained through structural identification with measurements before retrofit actions (Case 2 in Table 4). EDMF and modified BMU provide updated bounds for parameter $K_{A,z}$, while traditional BMU provides an informed (inaccurate) marginal PDF of $K_{A,z}$.

3.1.6. Comparison of Computational Cost

While EDMF and modified BMU provide compatible model updating, EDMF with grid sampling is more efficient than modified BMU in incorporating changing uncertainty definitions, such as changes to combined uncertainty estimation and prior parameter distributions.

Grid-sampling based model instances for EDMF were simulated using parallel computing. For the pre-retrofit scenario (Scenario 1 in Table 4), a grid of 273 model instances (IMS) was discretized into four groups and each group was simulated with a different core. Predictions from model instances were evaluated for compatibility with measurements using EDMF to obtain a CMS. Pre-retrofit, the first estimate of uncertainty without including model bias (Scenario 1 in Table 4), falsified all 273 model instances. As discussed in Section 3.1.1, this provides diagnostics to re-evaluate uncertainties and include model bias. Including model bias (Scenario 2 in Table 4), 127 model instances from 273 model instances were accepted into the CMS (see Figure 5a). Post-retrofit, no model instance from the IMS were found to be compatible with the new set of measurements (Scenario 3 in Table 4). This provides diagnostics to re-evaluate the prior distribution of model parameter $K_{A,z}$, as explained in Section 3.1.3. Due to change in prior distribution of model parameter $K_{A,z}$, 168 new model instances were added to the IMS, as shown in Figure 8 (Scenario 4 in Table 4). Only the additional model instances were evaluated again to obtain a new CMS using EDMF. CMS after changing prior distribution of model parameter $K_{A,z}$ contains 40 model instances. Thus, using EDMF with grid sampling helps save significant computational cost when iterative evaluations of uncertainty conditions and prior distribution of model parameters are required.

Modified BMU and traditional BMU were carried out in this study using MCMC sampling without parallel computing. MCMC sampling and other adaptive sampling strategies utilize assumptions of the posterior form of the solution space to improve efficiency of sampling. However, changes to prior distributions or uncertainty estimations (likelihood function) alter the solution space and, thus, require a complete re-start. This significantly increases computational cost when data interpretation has to be carried out iteratively. A comparison of the cumulative computational cost incurred in applying these methodologies for multiple scenarios of identification is shown in Figure 9.

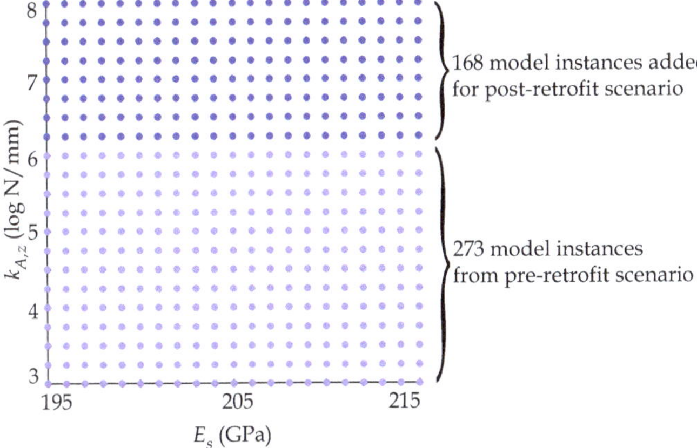

Figure 8. A change in prior distribution size requires simulation of only additional model instances to evaluate their compatibility with measurements when using EDMF.

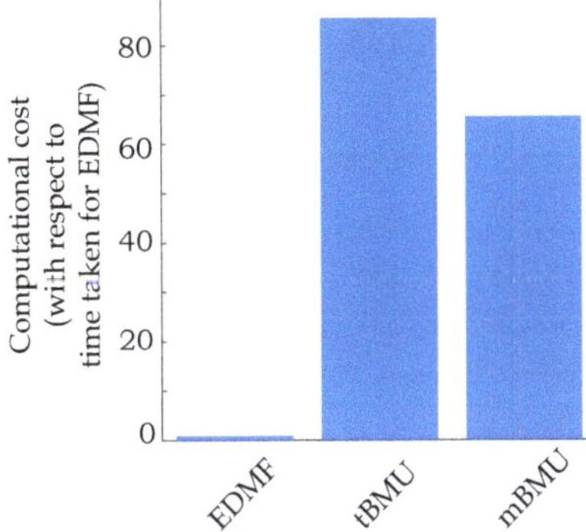

Figure 9. Comparison of computational cost. The simulations were carried out on a Intel(R) Xeon(R) CPU X5650 @2.67GHz processor with 24 cores.

Figure 9 shows the cumulative computational cost incurred related to the three probabilistic data-interpretation methodologies. As EDMF with grid sampling does not require a complete re-start between iterations of identification and can be carried out using parallel computing, it incurs the lowest computational cost. Modified and traditional BMU incur much higher computational costs due to repeated re-starts that increase evaluations required using computationally expensive physics-based models.

A key aspect that is not included in Figure 9 is the computational cost incurred in appropriately carrying out MCMC sampling for modified and traditional BMU. For MCMC sampling, the computational cost depends upon user-defined step-sizes, which affect acceptance rate and number of accepted samples to be simulated. If after obtaining a pre-defined number of samples from the joint posterior PDF, the variance of the PDF is not consistent (convergence criteria), then sampling has to be re-started. User-defined inputs of MCMC sampling are dependent upon the solution space and vary between cases. Their values have to be determined based on heuristics, following a trial-and-error process. Each iteration of trial-and-error adds to the computational cost and makes their application in practice challenging. Although grid sampling with EDMF is computationally expensive, a parallel-computing environment can reduce computation time efficiently for a small number of parameters. In addition, grid sampling is capable of transparently and efficiently incorporating changes that make data interpretation an iterative task, as samples are independent of likelihood functions and other assumptions.

3.2. Crêt de l'Anneau Bridge, Switzerland

In this study, the Crêt de l'Anneau Bridge, shown in Figure 10, was investigated using deflection measurements recorded during a load test. The Crêt de l'Anneau Bridge, built in 1969, is part of Route de la Promenade close to Neuchatel (Switzerland). Deflection measurements collected during a load test on this bridge were utilized to updated a FE model of the bridge.

Figure 10. Crêt de l'Anneau Bridge near Neuchatel (Switzerland).

The Crêt de l'Anneau Bridge is a steel-concrete bridge with nine spans, as shown in Figure 11a. The first and last spans, noted in the figure as Spans 0 and 8, connect the bridge to the roadway. Each of the bridge spans, denoted I to VII in Figure 11, have a length of 25.6 m. The bridge is curved, as shown in Figure 11a, and the total length of the bridge along the inner arc is 195 m. In the bridge, 5 m after support of each span, there is a Gerber joint (Figure 11a). Measurements were carried out with deflection sensors at the middle of Spans II and IV, for which cross-sections and sensor locations are shown in Figure 11b. Deflections were recorded under a load test, during which a truck of 40 ton was driven over the bridge in a traffic lane (Neuchatel to Travers direction) at 10 km/h. Let the sensors in Span II be numbered S1–S5 and Span IV be numbered S6–S10. Data from sensors S1–S5 when the truck is at the middle of Span II and S6–S10 when the truck is at middle of Span IV were utilized for model updating.

To interpret measurements recorded during the load test, a FE model of the bridge was developed in Ansys [85]. In the FE model, the concrete deck, the two main steel girders and intermediate transversal beams were modeled using shell elements. The concrete deck, as shown in Figure 11b, has a complex geometry, which was simplified in the model to have a constant thickness. The stiffness of Gerber joints and connectors between deck and girder in longitudinal (in direction of traffic) and transversal (perpendicular to direction of traffic) directions were parameterized using zero-length linear spring elements. The stiffness of supports (excluding those at ends of special spans, totaling eight supports) were parameterized in vertical and longitudinal directions. The concrete deck was modeled as homogeneous with the material model defined by the Young's modulus of concrete, which was included as a parameter in the FE model. The Young's modulus of steel was also included in the FE model as a parameter. The prior distributions of parameters included in the FE model are shown in Table 5.

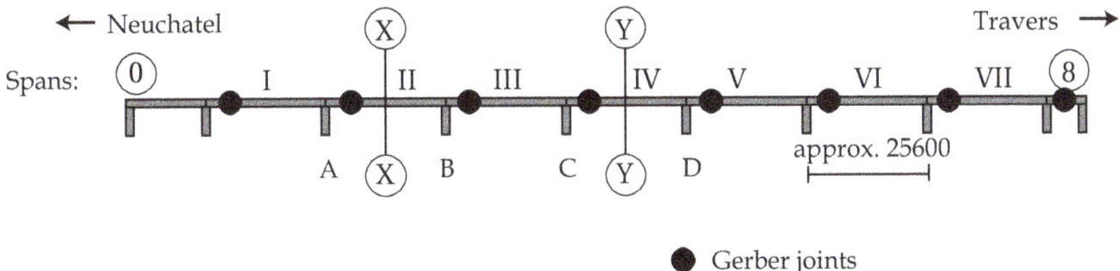

(a) Bridge elevation (Span 0 and 8 are special spans that connect the bridge to the road.)

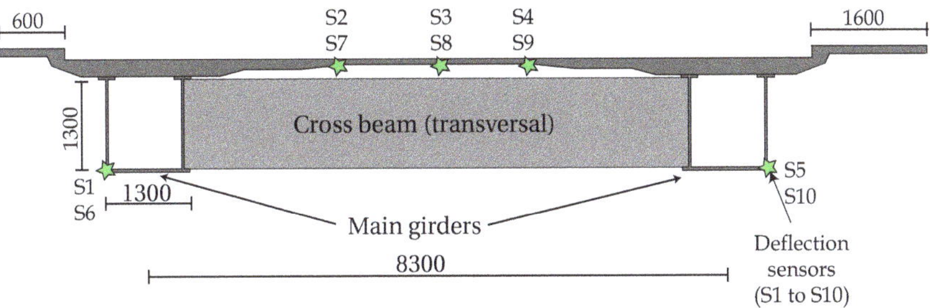

(b) Cross-section of a typical span (X-X or Y-Y)

Figure 11. (**a**) Elevation of Crêt de l'Anneau Bridge; and (**b**) cross-section of a typical span, showing the location of deflection sensors placed on Spans II and IV.

Table 5. Parameters included in the FE model and their prior probability distributions.

Description	Units	Distribution
Stiffness of supports (longitudinal)	log N/mm	U(3.5, 5.0)
Stiffness of supports (vertical)	log N/mm	U(3.5, 5.5)
Young's modulus of steel	GPa	U(190, 220)
Young's modulus of concrete	GPa	U(30, 50)
Gerber joint (longitudinal)	log N/mm	U(4.0, 6.0)
Deck to girder connection (longitudinal)	log N/mm	U(4.0, 5.5)

After a preliminary sensitivity analysis, six primary parameters were chosen for identification: stiffness of deck-to-girder connection, vertical support stiffness at supports of Spans II and IV (see Figure 11) and horizontal connection stiffness at the Gerber joints. The primary parameters and the corresponding initial parameter ranges are reported in Table 6. Ten equidistant samples were drawn from each primary parameter to obtain an initial grid of parameter combinations with 10^6 instances.

Parameters that were not retained as primary parameters to identify, such as Young's modulus of reinforced concrete and steel as well as rotational and horizontal stiffness of supports, were considered secondary parameters. Additional uncertainty arising from omitting these parameters was estimated using the FE model with Monte-Carlo sampling (see Table 7).

Evaluating a large parameter space (six parameters) using a computationally expensive FE model is time consuming. Thus, surrogate models were developed to predict response at sensor locations. The surrogate-modeling strategy employed here is Gaussian process regression, which provides precise surrogate models that are trained using a dataset generated with the FE model, wherein the primary parameters are varied within the bounds of their prior distribution. Uncertainty from use of surrogate models was determined using a second validation dataset. The uncertainty arising from using surrogate models was quantified as uniform, as shown in Table 7.

Table 6. Primary parameters with their parameter ranges (prior distributions) for the Crêt-de-l'Anneau bridge.

Parameter	Description	Units	Range
θ_1	Deck-to-girder connection stiffness (longitudinal)	log (N/mm)	U(4.0,5.5)
θ_2	Vertical stiffness support A	log (N/mm)	U(3.5, 5.5)
θ_3	Vertical stiffness support B	log (N/mm)	U(3.5, 5.5)
θ_4	Vertical stiffness support C	log (N/mm)	U(3.5, 5.5)
θ_5	Vertical stiffness support D	log (N/mm)	U(3.5, 5.5)
θ_6	Gerber joint stiffness (longitudinal)	log (N/mm)	U(4.0, 6.0)

Table 7. Secondary (%) and surrogate modeling (mm) uncertainty at each sensor location.

Source	Distribution
Model bias (%)	U(−5, 15)
Secondary parameters (%)	U(−1.5, 0.5)
Surrogate model (mm)	U(−0.1, 0.1)
Measurement (mm)	M(0, 0.15)

Other sources of uncertainty affecting identification are also shown in Table 7. Model bias arises from assumptions made during model development such as geometry of the concrete deck and supports.

Three scenarios for data-interpretation were performed using data from the two load tests to highlight strengths and shortcomings in practical applications of four data-interpretation methodologies:

- **Scenario 1**: Deflection measurements at five locations (S1–S5) were used and model uncertainty was ignored.
- **Scenario 2**: Deflection measurements at five locations (S1–S5) were used and model uncertainty was taken into account.
- **Scenario 3**: Deflection-measurements at 10 locations were used and model uncertainty was taken into account.

3.2.1. Structural Identification (Scenario 1: Ignoring Model Bias)

The first data-interpretation scenario involved deflection measurements from five sensors distributed in the transverse direction of one longitudinal location at Span II (see Figure 11). In this scenario, model uncertainties (model bias and surrogate-model uncertainty) were ignored when deriving combined uncertainties. In other words, combined uncertainty was under-estimated.

Using EDMF, all 10^6 parameter combinations were falsified as they produce residuals between measured and predicted deflections that do not comply with Equation (6) for all five measured locations. This indicates that either choice of model parameters and their ranges is erroneous or that uncertainty is under-estimated, as is the case here. In a similar way, modified BMU failed to find a starting point for 2000 randomly selected parameter combinations. Both methods led to the same conclusion that model predictions are incompatible with measurements, given the estimated combined uncertainties. However, modified BMU reached the conclusion with shorter simulation time.

Although uncertainties were under-estimated, traditional BMU provided updated parameter values. Unlike EDMF and modified BMU, traditional BMU does not involve thresholds and, thus, no strict delimitation of acceptance and rejection regions exists. Updated parameter values for traditional BMU are reported in Table 8. When comparing maximum a-posteriori (MAP) estimate with initial parameter values (see Table 6), vertical stiffness of supports 3 and 4 were estimated to be high while the one of supports 1 and 2 was intermediate. The 95th-percentile bounds on parameter marginals indicate that parameter uncertainty remains high for stiffness of supports 1 and 2, not located near the measured Span 4. Residual minimization carried out using grid sampling provided a single parameter instance as optimal, which is reported in Table 8.

Table 8. Updated parameter ranges (posterior distributions) for Scenario 1. mBMU failed to find a starting point, while EDMF falsified the entire initial model set. MAP refers to the maximum a-posteriori estimate.

Parameter	RM	tBMU	mBMU	EDMF
θ_1	5.5	$[5.1, 5.5]$, MAP = 5.5	-	-
θ_2	5.5	$[3.6, 5.4]$, MAP = 4.8	-	-
θ_3	5.5	$[3.7, 5.4]$, MAP = 4.8	-	-
θ_4	5.5	$[4.8, 5.5]$, MAP = 5.4	-	-
θ_5	5.5	$[4.9, 5.5]$, MAP = 5.4	-	-
θ_6	5.8	$[4.8, 6.0]$, MAP = 5.7	-	-

A powerful tool to assess accuracy and precision of parameter-updating results is leave-one-out cross validation, as discussed in Section 2.2. Using MCMC sampling, the updated parameter distributions were conditioned upon the likelihood function and the measurements. However, when leaving out a sensor, the dimensionality of the likelihood function changed: instead of five dimensions, it only had four dimensions. Thus, when sequentially leaving out all sensors locations, five new runs of traditional BMU needed to be performed, which increased computation time significantly. The results in terms of accuracy and average precision of traditional BMU over all five sensors are reported in Table 9. While high precision was achieved (resulting from low uncertainty values), accuracy was not validated for any sensor location using leave-one-out cross validation and 95th-percentile bounds of predictions. This indicates as well that uncertainties were under-estimated or that assumptions of the likelihood function, such as no correlation, were not appropriate. However, to reach to this conclusion, important simulation time was needed. Residual minimization, which provides a single optimal parameter instance, did not provide accurate deflection predictions at the sensor left-out for three out of five leave-one-out cases. A single parameter instance was precise, with a ϕ value of 1. The accuracy and precision for residual minimization is reported in Table 9.

Table 9. Accuracy and precision established using a leave-one-out cross-validation approach. For mBMU and EMDF, the entire model class was rejected and, thus, no updated parameter values could be validated. For tBMU, absence of accuracy indicates that uncertainties were mis-evaluated.

Leave-One-Out Cross-Validation	RM	tBMU	mBMU	EDMF
Accuracy	No	No	-	-
Precision	1	0.96	-	-

3.2.2. Structural Identification (Scenario 2: Five Measurement Locations)

While still involving deflection measurements at five locations, Scenario 2 implicitly took into account model uncertainty. Despite uncertainties still being centered on 0 for traditional BMU, the increase in variance translated to a change in MAP estimates, as can be seen by comparing updated values of Scenario 2, which are reported in Table 10, with updated values of Scenario 1 (Table 8). Residual minimization results do not change from Scenario 1 as uncertainties in identification were not considered in search for optimal parameter values.

EDMF and modified BMU did not provide informed posterior distributions and, thus, all values were considered equally likely. As can be seen from the identified values (Table 10), parameter uncertainty for vertical of supports 1 and 2, θ_2 and θ_3, was not reduced, as measurements were taken at Span 4, which is far from supports 1 and 2. Updated results of modified BMU and EDMF were equivalent, which underlines compatibility of the two data-interpretation methodologies. Grid sampling used for EDMF ensured that the complete parameter space was explored and, thus, updated range for parameter θ_6 was larger than for modified BMU.

Table 10. Updated parameter ranges (posterior distributions) for Scenario 2.

Parameter	RM	tBMU	mBMU	EDMF
θ_1	5.5	[4.9, 5.5], MAP = 5.4	[4.8, 5.5]	[4.8, 5.5]
θ_2	5.5	[3.5, 5.2], MAP = 4.0	[3.5, 5.5]	[3.5, 5.5]
θ_3	5.5	[3.5, 5.4], MAP = 4.6	[3.5, 5.5]	[3.5, 5.5]
θ_4	5.5	[4.6, 5.5], MAP = 5.4	[4.4, 5.5]	[4.4, 5.5]
θ_5	5.5	[4.7, 5.5], MAP = 5.4	[4.6, 5.5]	[4.6, 5.5]
θ_6	5.8	[4.3, 6.0], MAP = 5.8	[4.3, 6.0]	[4.0, 6.0]

Leave-one-out cross validation, when performed over all measured locations allows engineers to assess accuracy and precision of model updating. This step reduces the risk of wrong parameter updating that can lead to wrong predictions when extrapolation is performed. Figure 12 contains leave-one-out cross validation for separately leaving out each of the five sensor locations (δ_1 to δ_5) for all three probabilistic data-interpretation methodologies. Predictions at left-out sensor locations indicate that updated model parameters led to accurate predictions for all five measurements when either EDMF or modified BMU was used. Again, the prediction ranges from EDMF and modified BMU were the same for all five measurements, which underlines that the two methodologies are equivalent in terms of parameter updating.

However, when traditional BMU was used, the measurement fell outside the 95th-percentile bounds for sensors 1 and 2. This indicates that estimated uncertainty values or distributions were not compatible with the model class. Thus, as shown in Table 11, accuracy was verified for EDMF and modifed BMU and not for traditional BMU with independent zero-mean likelihood functions. However, precision was higher when using traditional BMU (see Table 11). In addition, MAPs that were provided by traditional BMU are biased with respect to measurements (see Figure 12).

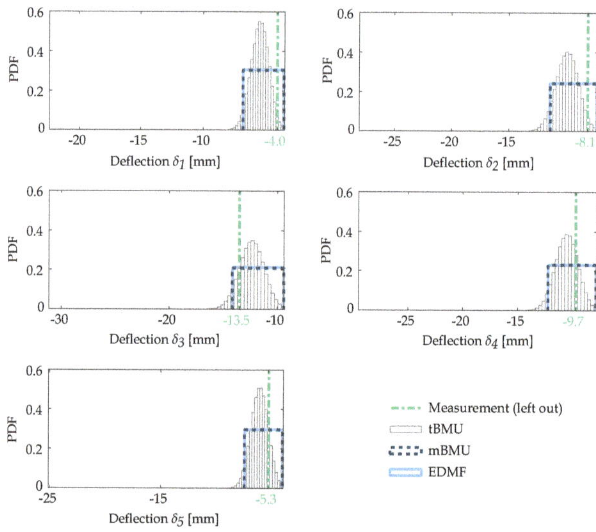

Figure 12. Leave-one-out cross validation for the five deflection measurements used in Scenario 2. mBMU and EDMF were accurate for all five measurements, while the measurements fell outside 95th-percentile bounds for deflection δ_1 in the case of tBMU.

Table 11. Accuracy and precision established using a leave-one-out cross-validation approach for Scenario 2.

Leave-One-Out Cross Validation	RM	tBMU	mBMU	EDMF
Accuracy	No	No	Yes	Yes
Precision	1	0.84	0.74	0.74

3.2.3. Structural Identification (Scenario 3: 10 Measurements Locations)

For the third scenario, the assumptions were the same as for previous Scenario 2 with deflection measurements at five additional locations (see Figure 11). As discussed in Section 3.2.4, additional measurements resulted in a higher dimensionality for the Bayesian likelihood function and, thus, MCMC simulations needed to be re-initiated. Grid-sampling-based application of EDMF offered more flexibility and only candidate models from Scenario 2 needed to be re-evaluated with respect to the new measurement locations, δ_6 to δ_{10}.

Updated parameter values for the four data-interpretation methodologies are provided in Table 12. For traditional BMU, MAP of parameters changed again with respect to Table 10, even for parameters that had little influence at newly added measurement locations. For all three methodologies, parameter uncertainty related to stiffness of supports 1 and 2, θ_2 and θ_3, was reduced. As the new measurements were taken in the span between these two supports, this result was expected.

In a similar way to the previous scenario, leave-one-out cross validation led to rejection of updated results for traditional BMU (see Table 13). Although standard deviations of the Gaussian likelihood functions were taken to be compatible with the combined uncertainty (used to derive thresholds for EDMF and modified BMU), the distribution was zero-mean and independence between measurement locations was assumed. Thus, updated parameter values may lead to inaccurate identification if these conditions were not met. Again, precision was higher for traditional BMU than for other methods. This was expected since EDMF and modified BMU sacrifice precision in order to be robust with respect to biased and correlated error sources. Residual minimization provides precise (single parameter instance) albeit inaccurate model updating as uncertainties affecting structural identification and correlation between measurement locations were not taken into consideration.

Table 12. Updated parameter ranges (posterior distributions) for Scenario 3 involving deflection measurements at ten locations.

Parameter	RM	tBMU	mBMU	EDMF
θ_1	5.5	[5.0, 5.5], MAP = 5.4	[5.0, 5.5]	[5.0, 5.5]
θ_2	5.5	[4.6, 5.5], MAP = 5.4	[4.6, 5.5]	[4.6, 5.5]
θ_3	5.5	[4.8, 5.5], MAP = 5.3	[4.8, 5.5]	[4.8, 5.5]
θ_4	5.5	[4.5, 5.5], MAP = 5.3	[4.4, 5.5]	[4.4, 5.5]
θ_5	5.5	[4.6, 5.5], MAP = 5.3	[4.5, 5.5]	[4.6, 5.5]
θ_6	5.8	[5.0, 6.0], MAP = 5.9	[4.8, 6.0]	[4.4, 6.0]

Table 13. Accuracy and precision established using a leave-one-out cross-validation approach for Scenario 3, involving ten deflection measurements. Precision is the mean value over the ten measurement locations, while accuracy needs to be validated over all measurement locations.

Leave-One-Out Cross Validation	RM	tBMU	mBMU	EDMF
Accuracy	No	No	Yes	Yes
Precision	1	0.85	0.82	0.81

3.2.4. Practical Aspects of Data Interpretation

Identification results presented in the previous sections are in accordance with previous findings [30–32]. A main aspect of this paper is to compare the three methodologies with respect to their compatibility with practical needs. The main practical aspect related to the this case study is computation time related to iterative data interpretation.

As stated above, data interpretation is a fundamentally iterative task. Information becomes available over time and assumptions, for instance regarding uncertainty sources, need to be re-evaluated frequently. This case study reflects these two aspects: while from the first to the second scenario uncertainties are re-evaluated, the third scenario adds additional measurements. Figure 13 gives the computation time for the three data-interpretation methodologies when performing the three scenarios iteratively. The computation times provided in Figure 13 were based on a Intel(R) Xeon(R) CPU E5-2670 v3 @2.30 GHz processor with up to 24 cores used in parallel. While computation time for EDMF can be divided by 24 using parallel computing (instances of grid sampling are independent and do not involve communication between parallel cores), only leave-one-out cross validation can be run in parallel for BMU applications, as each left-out sensor can be simulated independently.

When using EDMF, performing leave-one-out cross validation is computationally efficient as it does not require additional simulations. Leaving out information or adding information (if already simulated) only requires to re-evaluate the threshold values, due to the Sidak correction. In a similar way, when using EDMF, changes in uncertainties (first iteration) do not require additional simulations of the physics-based model, only threshold values are re-calculated. Thus, even if grid sampling is computationally expensive up front (exponential complexity with respect to number of divisions and number of parameters), it offers high flexibility for exploration of data-interpretation results. In addition, EDMF involves validating Equation (7) for all measurements. Thus, when adding new measurements (Scenario 3), only candidate models need to be re-simulated, which decreases computation time (see Figure 4e).

When Bayesian approaches are used, changes in uncertainties (likelihood function) and additional measurement require re-simulations using MCMC sampling. Thus, even if adaptive sampling provides an opportunity to reduce computational complexity of parameter-space exploration, every subsequent change results in new simulations of structural behavior. However, for very high number of parameters, which are typically encountered in structures with multiple parallel loading paths, MCMC sampling outperforms grid sampling. Unless many data are available, such structures are usually unidentifiable and, thus, direct measurements of causes, rather than effects, are required.

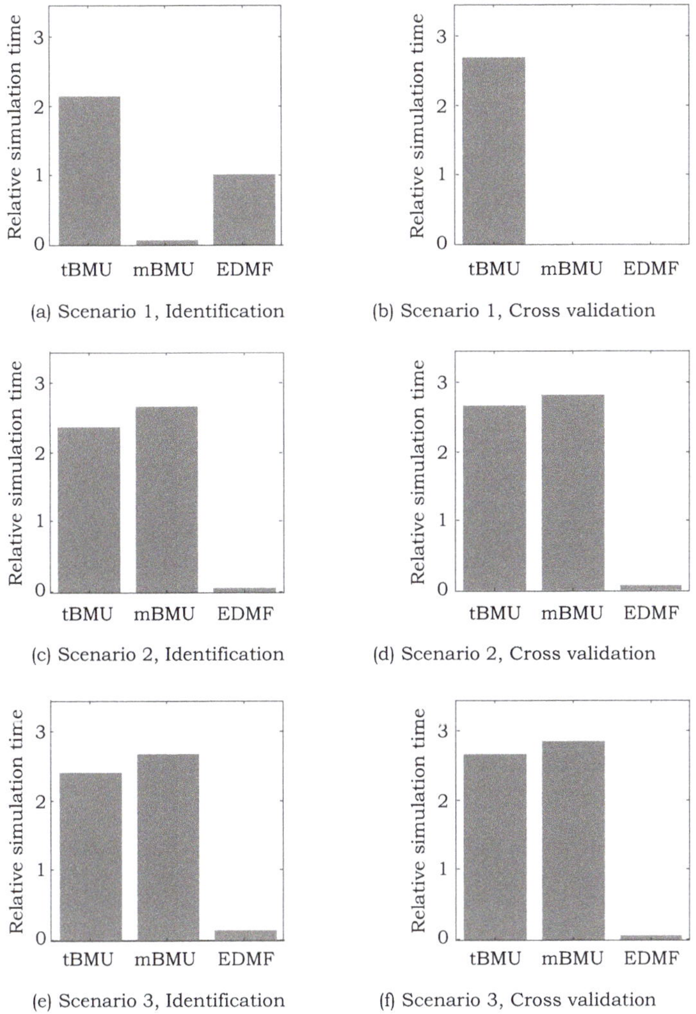

(a) Scenario 1, Identification (b) Scenario 1, Cross validation

(c) Scenario 2, Identification (d) Scenario 2, Cross validation

(e) Scenario 3, Identification (f) Scenario 3, Cross validation

Figure 13. Comparison of computation times for the three probabilistic data-interpretation methodologies. Computation time is for parallel computing using up to 24 cores and is presented in relative time with respect to simulation time for EDMF in Scenario 1 (A).

4. Discussion of Results

4.1. Ponneri Bridge Case Study

In Table 14, a summary of accuracy achieved using the four data-interpretation methodologies for various scenarios of structural identification of Ponneri Bridge is shown. EDMF and modified BMU provided accurate model updating for all scenarios. Residual minimization lacked accuracy as it did not take into account the model bias for parameter estimation. Traditional BMU was accurate for only one of the four scenarios. EDMF had additional advantages over modified BMU due to the sampling strategy employed that allows for computationally efficient and transparent inclusion of changes to uncertainty definitions and prior parameter distributions.

Table 14. Summary of the accuracy of structural identification scenarios (see Table 4) evaluated for the Ponneri Bridge. mBMU is modified BMU and tBMU is traditional BMU. Checkmarks imply accurate identification and crosses imply inaccurate identification.

Case	Scenario	Description	RM	EDMF	mBMU	tBMU
1	Before retrofit	Without model bias	✗	✓	✓	✗
2		With model bias	✗	✓	✓	✗
3	After retrofit	Without re-evaluating prior PDFs	✗	✓	✓	✗
4		After re-evaluating prior PDFs	✗	✓	✓	✓

4.2. *Crêt de l'Anneau Bridge Case-Study*

For the Crêt-de-l'Anneau bridge, real measurements were used for model updating. Thus, unlike the Ponneri Bridge, real parameter values were unknown and validation of updating results can only be obtained using leave-one-out cross validation. Table 15 contains a summary of accuracy achieved using the four data-interpretation methodologies.

Figure 14 provides updated prediction results at sensors S7 and S8 before and after deflection measurements were performed in the span containing these sensors. Modified BMU and EDMF were equivalent in terms of updating results and, thus, provided the same uninformed prediction ranges. Traditional BMU uses an informed likelihood function and thus points towards a value that has higher probabilities than others, the MAP. As can be observed in Figure 14, this value is often biased from the measured value. Predictions made using MAP values are biased and might lead to potentially un-conservative results, as can be seen in Figure 14D. In addition, adding more measurements may increase precision, not accuracy.

The typical situation of bridge case studies is that knowledge of real parameter values is not available. Thus, validation of identification is most efficiently carried out with leave-one-out cross-validation. Moreover, new information from additional sensors leads to iterations of data-interpretation. These iterations follow the steps shown in Figure 1.

Table 15. Summary of the accuracy of structural identification scenarios evaluated for the Crêt-de-l'Anneau Bridge. mBMU is modified BMU and tBMU is traditional BMU. Checkmarks imply accurate identification or model-class rejection and crosses imply inaccurate identification based on leave-one-out cross validation.

Scenario	Description	RM	EDMF	mBMU	tBMU
1	Without model bias (deflection at 5 locations)	✗	✓	✓	✗
2	With model bias (deflection at 5 locations)	✗	✓	✓	✗
3	With model bias (deflection at 10 locations)	✗	✓	✓	✗

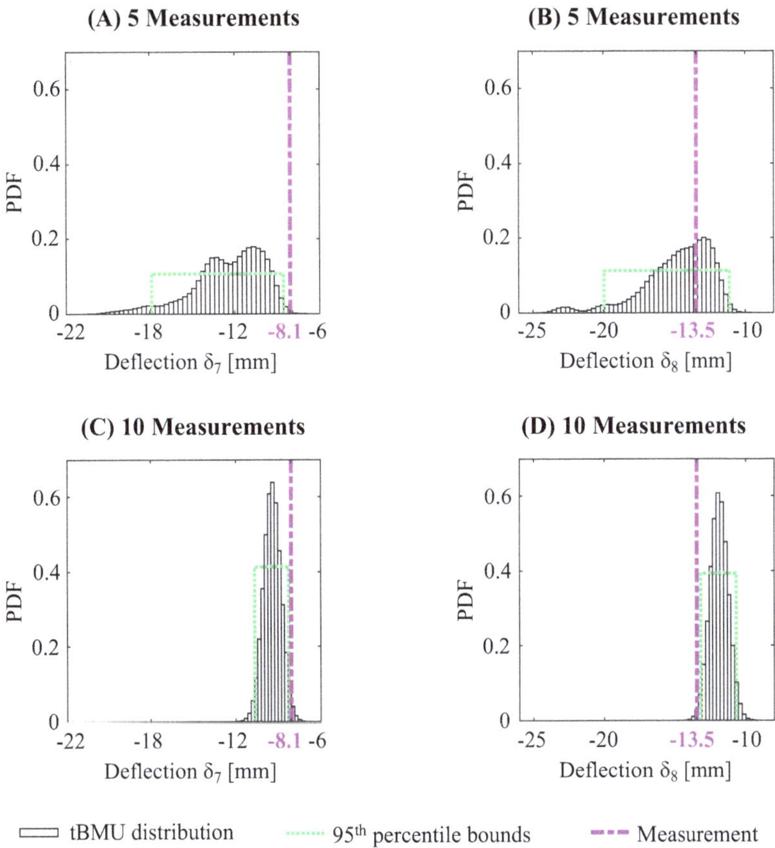

Figure 14. Leave-one-out cross validation for sensors 7 (**A**,**C**) and 8 (**B**,**D**) before (**A**,**B**) and after (**C**,**D**) including measurements at the same span in structural identification for the Cret-de-l'Anneau bridge. Although inclusion of measurements from the same span increases precision, 95th-percentile bounds are not compatible with measurements in both cases. In addition, displacement is overestimated for sensor 7 (**C**) and underestimated for sensor 8 (**D**), which shows that results are not always conservative.

5. Conclusions

In this paper, practical challenges related to application of three data-interpretation methodologies are evaluated with the use of two full-scale case studies. Conclusions are as follows:

- EDMF incorporates new information such as changes to uncertainty definitions and additional measurements iteratively. Bayesian model-updating methodologies and residual minimization must be restarted each time.
- EDMF is computationally more efficient than Bayesian model updating methodologies in an iterative data-interpretation framework, especially when grid sampling is used in combination with parallel computing.
- EDMF and modified BMU provide updated bounds on parameter values, which is more interpretable for practicing engineers than posterior parameter distributions that are obtained using traditional BMU.

- Residual minimization provides single optimal parameter values and, while this is compellingly useful in practice, it is not accurate in the presence of the biased uncertainties that are common in engineering modeling.
- EDMF involves a procedure that is more compatible with typical engineering procedures. For example, it is customary to define target reliability levels at the beginning. EDMF follows this procedure, whereas Bayesian approaches leave this to the end.
- Accuracy is assessed using leave-one-out cross-validation, which is computationally inexpensive when EDMF with grid sampling is used and computationally expensive when traditional BMU methodology is used.

The studies that are described in this paper illustrate the advantages of using EDMF for practical engineering diagnosis and prediction tasks that are supported by measurements. Future work involves a user-centric study to understand better the challenges that have to be addressed to enable use of EDMF in practice.

Author Contributions: S.G.S.P. and Y.R. conducted analyses of the case studies, demonstrating advantages in use of EDMF in an iterative framework for data-interpretation. I.F.C.S. was actively involved in developing and adapting the data-interpretation methodologies. All authors were in involved in writing the paper, and reviewed and accepted the final version.

Funding: This work was funded by the Swiss National Science Foundation under contract No. 200020-169026 and Singapore-ETH Centre (SEC) under contract No. FI 370074011-370074016.

Acknowledgments: The authors acknowledge R. Wegmann for development of the model of Crêt de l'Anneau bridge and B. Raphael for providing design drawings of the Ponneri Bridge.

Conflicts of Interest: The authors declare no conflict of interest.

References

1. World Economic Forum; The Boston Consulting Group. *Shaping the Future of Construction: A Breakthrough in Mindset and Technology*; World Economic Forum: Cologny, Switzerland, 2016.
2. World Economic Forum. *Strategic Infrastructure, Steps to Operate and Maintain Infrastructure Efficiently and Effectively*; World Economic Forum: Cologny, Switzerland, 2014.
3. Brühwiler, E. Extending the service life of Swiss bridges of cultural value. *Proc. Inst. Civ. Eng. Eng. Hist. Herit.* **2012**, *165*, 235–240. [CrossRef]
4. Smith, I.F.C. Studies of Sensor-Data Interpretation for Asset Management of the Built Environment. *Front. Built Environ.* **2016**, *2*, 8. [CrossRef]
5. World Economic Forum; The Boston Consulting Group. *Shaping the Future of Construction: Inspiring Innovators Redefine the Industry*; World Economic Forum: Cologny, Switzerland, 2017.
6. Lynch, J.P.; Loh, K.J. A summary review of wireless sensors and sensor networks for structural health monitoring. *Shock Vib. Dig.* **2006**, *38*, 91–130. [CrossRef]
7. Taylor, S.G.; Raby, E.Y.; Farinholt, K.M.; Park, G.; Todd, M.D. Active-sensing platform for structural health monitoring: Development and deployment. *Struct. Health Monit.* **2016**, *15*, 413–422. [CrossRef]
8. Frangopol, D.M.; Soliman, M. Life-cycle of structural systems: Recent achievements and future directions. *Struct. Infrastruct. Eng.* **2016**, *12*, 1–20. [CrossRef]
9. Der Kiureghian, A. Analysis of structural reliability under parameter uncertainties. *Probab. Eng. Mech.* **2008**, *23*, 351–358. [CrossRef]
10. Jiang, X.; Mahadevan, S. Bayesian validation assessment of multivariate computational models. *J. Appl. Stat.* **2008**, *35*, 49–65. [CrossRef]
11. Mottershead, J.E.; Friswell, M. Model updating in structural dynamics: a survey. *J. Sound Vib.* **1993**, *167*, 347–375. [CrossRef]
12. Soize, C. Stochastic models of uncertainties in computational structural dynamics and structural acoustics. In *Nondeterministic Mechanics*; Springer: Berlin, Germany, 2012; pp. 61–113.
13. Soize, C. Generalized probabilistic approach of uncertainties in computational dynamics using random matrices and polynomial chaos decompositions. *Int. J. Numer. Methods Eng.* **2010**, *81*, 939–970. [CrossRef]

14. Görl, E.; Link, M. Damage identification using changes of eigenfrequencies and mode shapes. *Mech. Syst. Signal Process.* **2003**, *17*, 103–110. [CrossRef]
15. Beck, J.L. Bayesian system identification based on probability logic. *Struct. Control Health Monit.* **2010**, *17*, 825–847. [CrossRef]
16. Cross, E.J.; Worden, K.; Farrar, C.R. Structural health monitoring for civil infrastructure. In *Health Assessment of Engineered Structures: Bridges, Buildings and Other Infrastructures*; World Scientific: Hackensack, NJ, USA, 2013; pp. 1–28.
17. Moon, F.; Catbas, N. Structural Identification of Constructed Systems. In *Structural Identification of Constructed Systems*; American Society of Civil Engineers: Reston, VA, USA, 2013; pp. 1–17. [CrossRef]
18. Sanayei, M.; Imbaro, G.R.; McClain, J.A.; Brown, L.C. Structural model updating using experimental static measurements. *J. Struct. Eng.* **1997**, *123*, 792–798. [CrossRef]
19. Beven, K.J. Uniqueness of place and process representations in hydrological modelling. *Hydrol. Earth Syst. Sci. Discuss.* **2000**, *4*, 203–213. [CrossRef]
20. Mottershead, J.E.; Link, M.; Friswell, M.I. The sensitivity method in finite element model updating: A tutorial. *Mech. Syst. Signal Process.* **2011**, *25*, 2275–2296. [CrossRef]
21. McFarland, J.; Mahadevan, S. Error and variability characterization in structural dynamics modeling. *Comput. Methods Appl. Mech. Eng.* **2008**, *197*, 2621–2631. [CrossRef]
22. McFarland, J.; Mahadevan, S. Multivariate significance testing and model calibration under uncertainty. *Comput. Methods Appl. Mech. Eng.* **2008**, *197*, 2467–2479. [CrossRef]
23. Rebba, R.; Mahadevan, S. Validation of models with multivariate output. *Reliab. Eng. Syst. Saf.* **2006**, *91*, 861–871. [CrossRef]
24. Beck, J.L.; Katafygiotis, L.S. Updating models and their uncertainties. I: Bayesian statistical framework. *J. Eng. Mech.* **1998**, *124*, 455–461. [CrossRef]
25. Kennedy, M.C.; O'Hagan, A. Bayesian calibration of computer models. *J. R. Stat. Soc. Ser. B* **2001**, *63*, 425–464. [CrossRef]
26. Brynjarsdóttir, J.; O'Hagan, A. Learning about physical parameters: The importance of model discrepancy. *Inverse Probl.* **2014**, *30*, 114007. [CrossRef]
27. Li, Y.; Xiao, F. Bayesian Update with Information Quality under the Framework of Evidence Theory. *Entropy* **2019**, *21*, 5. [CrossRef]
28. Simoen, E.; Papadimitriou, C.; Lombaert, G. On prediction error correlation in Bayesian model updating. *J. Sound Vib.* **2013**, *332*, 4136–4152. [CrossRef]
29. Goulet, J.A.; Smith, I.F.C. Structural identification with systematic errors and unknown uncertainty dependencies. *Comput. Struct.* **2013**, *128*, 251–258. [CrossRef]
30. Pasquier, R.; Smith, I.F. Robust system identification and model predictions in the presence of systematic uncertainty. *Adv. Eng. Inform.* **2015**, *29*, 1096–1109. [CrossRef]
31. Pai, S.G.; Nussbaumer, A.; Smith, I.F. Comparing structural identification methodologies for fatigue life prediction of a highway bridge. *Front. Built Environ.* **2018**, *3*, 73. [CrossRef]
32. Reuland, Y.; Lestuzzi, P.; Smith, I.F. Data-interpretation methodologies for non-linear earthquake response predictions of damaged structures. *Front. Built Environ.* **2017**, *3*, 43. [CrossRef]
33. Pasquier, R.; Smith, I.F.C. Iterative structural identification framework for evaluation of existing structures. *Eng. Struct.* **2016**, *106*, 179–194. [CrossRef]
34. Reuland, Y.; Lestuzzi, P.; Smith, I.F. A model-based data-interpretation framework for post-earthquake building assessment with scarce measurement data. *Soil Dyn. Earthq. Eng.* **2019**, *116*, 253–263. [CrossRef]
35. Zhang, Y.; O'Connor, S.M.; van der Linden, G.W.; Prakash, A.; Lynch, J.P. SenStore: A scalable cyberinfrastructure platform for implementation of data-to-decision frameworks for infrastructure health management. *J. Comput. Civ. Eng.* **2016**, *30*, 04016012. [CrossRef]
36. Worden, K.; Farrar, C.R.; Manson, G.; Park, G. The fundamental axioms of structural health monitoring. In *Proceedings of the Royal Society of London A: Mathematical, Physical and Engineering Sciences*; The Royal Society: London, UK, 2007; Volume 463, pp. 1639–1664.
37. Pavlovskis, M.; Antucheviciene, J.; Migilinskas, D. Application of MCDM and BIM for evaluation of asset redevelopment solutions. *Stud. Inform. Control* **2016**, *25*, 293–302. [CrossRef]
38. Vinogradova, I.; Podvezko, V.; Zavadskas, E. The recalculation of the weights of criteria in MCDM methods using the bayes approach. *Symmetry* **2018**, *10*, 205. [CrossRef]

39. Kaganova, O.; Telgarsky, J. Management of capital assets by local governments: An assessment and benchmarking survey. *Int. J. Strateg. Prop. Manag.* **2018**, *22*, 143–156. [CrossRef]
40. Re Cecconi, F.M.N.; Dejaco, M.C. Measuring the performance of assets: a review of the Facility Condition Index. *Int. J. Strateg. Prop. Manag.* **2018**, *23*, 187–196. [CrossRef]
41. Ljung, L. Perspectives on system identification. *Annu. Rev. Control* **2010**, *34*, 1–12. [CrossRef]
42. Chang, C.C.; Chang, T.; Xu, Y. Adaptive neural networks for model updating of structures. *Smart Mater. Struct.* **2000**, *9*, 59. [CrossRef]
43. Kuok, S.C.; Yuen, K.V. Investigation of modal identification and modal identifiability of a cable-stayed bridge with Bayesian framework. *Smart Struct. Syst.* **2016**, *17*, 445–470. [CrossRef]
44. Behmanesh, I.; Moaveni, B. Accounting for environmental variability, modeling errors, and parameter estimation uncertainties in structural identification. *J. Sound Vib.* **2016**, *374*. [CrossRef]
45. Friswell, M.; Penny, J.; Garvey, S. A combined genetic and eigensensitivity algorithm for the location of damage in structures. *Comput. Struct.* **1998**, *69*, 547–556. [CrossRef]
46. Ding, Z.; Huang, M.; Lu, Z. Structural damage detection using artificial bee colony algorithm with hybrid search strategy. *Swarm Evolut. Comput.* **2016**, *28*, 1–13. [CrossRef]
47. Gökdağ, H. Comparison of ABC, CPSO, DE and GA Algorithms in FRF Based Structural Damage Identification. *Mater. Test.* **2013**, *55*, 796–802. [CrossRef]
48. Gökdağ, H.; Yildiz, A.R. Structural damage detection using modal parameters and particle swarm optimization. *Mater. Test.* **2012**, *54*, 416–420. [CrossRef]
49. Majumdar, A.; Maiti, D.K.; Maity, D. Damage assessment of truss structures from changes in natural frequencies using ant colony optimization. *Appl. Math. Comput.* **2012**, *218*, 9759–9772. [CrossRef]
50. Beck, J.L.; Au, S.K. Bayesian updating of structural models and reliability using Markov chain Monte Carlo simulation. *J. Eng. Mech.* **2002**, *128*, 380–391. [CrossRef]
51. Ching, J.; Chen, Y.C. Transitional Markov chain Monte Carlo method for Bayesian model updating, model class selection, and model averaging. *J. Eng. Mech.* **2007**, *133*, 816–832. [CrossRef]
52. Boulkaibet, I.; Mthembu, L.; Marwala, T.; Friswell, M.I.; Adhikari, S. Finite element model updating using the shadow hybrid Monte Carlo technique. *Mech. Syst. Signal Process.* **2015**, *52*, 115–132. [CrossRef]
53. Dubbs, N.; Moon, F. Comparison and implementation of multiple model structural identification methods. *J. Struct. Eng.* **2015**, *141*, 04015042. [CrossRef]
54. Proverbio, M.; Costa, A.; Smith, I.F.C. Adaptive Sampling Methodology for Structural Identification Using Radial-Basis Functions. *J. Comput. Civ. Eng.* **2018**, *32*, 1–17. [CrossRef]
55. Robert-Nicoud, Y.; Raphael, B.; Smith, I. System Identification through Model Composition and Stochastic Search. *J. Comput. Civ. Eng.* **2005**, *19*, 239–247. [CrossRef]
56. Schwer, L.E.; Mair, H.U.; Crane, R.L. Guide for verification and validation in computational solid mechanics. *Am. Soc. Mech. Eng.* **2006**, *10*, 2006.
57. Alvin, K. Finite element model update via Bayesian estimation and minimization of dynamic residuals. *AIAA J.* **1997**, *35*, 879–886. [CrossRef]
58. Katafygiotis, L.S.; Beck, J.L. Updating models and their uncertainties. II: Model identifiability. *J. Eng. Mech.* **1998**, *124*, 463–467. [CrossRef]
59. Katafygiotis, L.S.; Papadimitriou, C.; Lam, H.F. A probabilistic approach to structural model updating. *Soil Dyn. Earthq. Eng.* **1998**, *17*, 495–507. [CrossRef]
60. Ching, J.; Beck, J.L. New Bayesian model updating algorithm applied to a structural health monitoring benchmark. *Struct. Health Monit.* **2004**, *3*, 313–332. [CrossRef]
61. Yuen, K.V.; Beck, J.L.; Katafygiotis, L.S. Efficient model updating and health monitoring methodology using incomplete modal data without mode matching. *Struct. Control Health Monit.* **2006**, *13*, 91–107. [CrossRef]
62. Muto, M.; Beck, J.L. Bayesian updating and model class selection for hysteretic structural models using stochastic simulation. *J. Vib. Control* **2008**, *14*, 7–34. [CrossRef]
63. Ntotsios, E.; Papadimitriou, C.; Panetsos, P.; Karaiskos, G.; Perros, K.; Perdikaris, P.C. Bridge health monitoring system based on vibration measurements. *Bull. Earthq. Eng.* **2009**, *7*, 469. [CrossRef]
64. Goller, B.; Schueller, G. Investigation of model uncertainties in Bayesian structural model updating. *J. Sound Vib.* **2011**, *330*, 6122–6136. [CrossRef] [PubMed]
65. Sohn, H.; Law, K.H. Bayesian probabilistic damage detection of a reinforced-concrete bridge column. *Earthq. Eng. Struct. Dyn.* **2000**, *29*, 1131–1152. [CrossRef]

66. Beck, J.L.; Au, S.K.; Vanik, M.W. Monitoring structural health using a probabilistic measure. *Comput.-Aided Civ. Infrastruct. Eng.* **2001**, *16*, 1–11. [CrossRef]
67. Tarantola, A. *Inverse Problem Theory and Methods for Model Parameter Estimation*; Society for Industrial and Applied Mathematics (SIAM): Philadelphia, PA, USA, 2005.
68. Popper, K. *The Logic of Scientific Discovery*; Routledge: Abingdon-on-Thames, UK, 1959.
69. Šidák, Z. Rectangular confidence regions for the means of multivariate normal distributions. *J. Am. Stat. Assoc.* **1967**, *62*, 626–633. [CrossRef]
70. Goulet, J.A.; Michel, C.; Smith, I.F.C. Hybrid probabilities and error-domain structural identification using ambient vibration monitoring. *Mech. Syst. Signal Process.* **2013**, *37*, 199–212. [CrossRef]
71. Goulet, J.A.; Coutu, S.; Smith, I.F.C. Model falsification diagnosis and sensor placement for leak detection in pressurized pipe networks. *Adv. Eng. Inform.* **2013**, *27*, 261–269. [CrossRef]
72. Moser, G.; Paal, S.G.; Smith, I.F. Performance comparison of reduced models for leak detection in water distribution networks. *Adv. Eng. Inform.* **2015**, *29*, 714–726. [CrossRef]
73. Vernay, D.G.; Raphael, B.; Smith, I.F.C. A model-based data-interpretation framework for improving wind predictions around buildings. *J. Wind Eng. Ind. Aerodyn.* **2015**, *145*, 219–228. [CrossRef]
74. Pasquier, R.; Goulet, J.A.; Acevedo, C.; Smith, I.F.C. Improving Fatigue Evaluations of Structures Using In-Service Behavior Measurement Data. *J. Bridge Eng.* **2014**, *19*, 4014045. [CrossRef]
75. Pasquier, R.; Angelo, L.D.; Goulet, J.A.; Acevedo, C.; Nussbaumer, A.; Smith, I.F.C. Measurement, Data Interpretation, and Uncertainty Propagation for Fatigue Assessments of Structures. *J. Bridge Eng.* **2016**, *21*. [CrossRef]
76. Pai, S.G.S.; Smith, I.F.C. Comparing Three Methodologies for System Identification and Prediction. In Proceedings of the 14th International Probabilistic Workshop, Ghent, Belgium, 22 December 2017; Caspeele, R., Taerwe, L., Proske, D., Eds.; Springer International Publishing: Berlin, Germany, 2017; pp. 81–95. [CrossRef]
77. Goulet, J.A.; Smith, I.F.C. Performance-driven measurement system design for structural identification. *J. Comput. Civ. Eng.* **2012**, *27*, 427–436. [CrossRef]
78. Goulet, J.A.; Smith, I.F.C. Predicting the usefulness of monitoring for identifying the behavior of structures. *J. Struct. Eng.* **2012**, *139*, 1716–1727. [CrossRef]
79. Papadopoulou, M.; Raphael, B.; Smith, I.F.C.; Sekhar, C. Optimal sensor placement for time-dependent systems: Application to wind studies around buildings. *J. Comput. Civ. Eng.* **2015**, *30*, 4015024. [CrossRef]
80. Papadopoulou, M.; Raphael, B.; Smith, I.F.; Sekhar, C. Evaluating predictive performance of sensor configurations in wind studies around buildings. *Adv. Eng. Inform.* **2016**, *30*, 127–142. [CrossRef]
81. Reuland, Y.; Lestuzzi, P.; Smith, I.F. Measurement-based support for post-earthquake assessment of buildings. *Struct. Infrastruct. Eng.* **2019**, *5*, 1–16. [CrossRef]
82. Sychterz, A.C.; Smith, I.F. Using dynamic measurements to detect and locate ruptured cables on a tensegrity structure. *Eng. Struct.* **2018**, *173*, 631–642. [CrossRef]
83. Reuland, Y.; Pai, S.G.; Drira, S.; Smith, I.F. Vibration-based occupant detection using a multiple-model approach. In Proceedings of the IMAC XXXV—Structural Dynamics Challenges in Next Generation Aerospace Systems, Garden Grove, CA, USA, 30 January–2 February 2017; Society for Experimental Mechanics (SEM): Bethel, CT, USA, 2017.
84. Kohavi, R. A study of cross-validation and bootstrap for accuracy estimation and model selection. In Proceedings of the International Joint Conference on Artificial Intelligence (IJCAI), Montreal, QC, Canada, 20–25 August 1995; Volume 14, pp. 1137–1145.
85. APDL. *Mechanical Applications Theory Reference*, 13th ed.; ANSYS Release 13.0; ANSYS Inc.: Canonsburg, PA, USA, 2010.
86. Vrouwenvelder, T. The JCSS probabilistic model code. *Struct. Saf.* **1997**, *19*, 245–251. [CrossRef]

Journal of
Sensor and
Actuator Networks

Article

Swarm-Based Parallel Control of Adjacent Irregular Buildings Considering Soil–Structure Interaction

Mohsen Azimi [1,*] and Asghar Molaei Yeznabad [2]

1 Department of Civil and Environmental Engineering, North Dakota State University, Fargo, ND 58105, USA
2 Department of Electrical Engineering, University of Mohaghegh Ardabili, Ardabil 56199-11367, Iran; asgarmolaei1996@gmail.com
* Correspondence: mohsen.azimi@ndsu.edu; Tel.: +1-510-859-7705

Received: 4 February 2020; Accepted: 28 March 2020; Published: 30 March 2020

Abstract: Seismic behavior of tall buildings depends upon the dynamic characteristics of the structure, as well as the base soil properties. To consider these factors, the equations of motion for a multi-story 3D building are developed to include irregularity and soil–structure interaction (SSI). Inspired by swarm intelligence in nature, a new control method, known as swarm-based parallel control (SPC), is proposed in this study to improve the seismic performance and minimize the pounding hazards, by sharing response data among the adjacent buildings at each floor level, using a wireless-sensors network (WSN). The response of individual buildings is investigated under historic earthquake loads, and the efficiencies of each different control method are compared. To verify the effectiveness of the proposed method, the numerical example of a 15-story, 3D building is modeled, and the responses are mitigated, using semi-actively controlled magnetorheological (MR) dampers employing the proposed control algorithm and fuzzy logic control (FLC), as well as the passive-on/off methods. The main discussion of this paper is the efficiency of the proposed SPC over the independent FLC during an event where one building is damaged or uncontrolled, and an active control based upon the linear quadratic regulator (LQR) is considered for the purpose of having a benchmark ideal result. Results indicate that in case of failure in the control system, as well as the damage in the structural elements, the proposed method can sense the damage in the building, and update the control forces in the other adjacent buildings, using the modified FLC, so as to avoid pounding by minimizing the responses.

Keywords: swarm-based parallel control (SPC); Internet of Things (IoT); soil–structure interaction (SSI); semi-active control; adjacent buildings

---

1. Introduction

Since the nature of an earthquake is its unpredictable characteristics, the traditional passive design approaches do not guarantee the minimum damages with economic designs [1–3]. Consequently, buildings and bridges are continuously getting smarter, using intelligent control devices, as well as state-of-the-art structural health monitoring technologies [4–6]. In the past decade, the vibration control of adjacent buildings under seismic and wind loads has gained significant attention. The most frequently proposed solution for controlling the adjacent building is to couple them with actuators or dampers, which involves the installation of the control devices, such as semi-active dampers between two buildings to improve the performances of both structures. New studies have also revealed the promising potentials of a special type of dampers, triangular-plate added damping and stiffness (TADAS), for seismic vibration control applications [7,8], which can be used for pounding hazard mitigation, as well.

A significant number of studies in the literature have used semi-active control algorithms, based on LQR [9], LQG [10] and FLC [11], to control magnetorheological (MR) dampers [12], and of particular interest are the optimal controls using metaheuristic and neural network algorithms [13–15]. Despite the advances, coupling adjacent buildings is not always the best solution, particularly when the dimensions, as well as the other dynamic properties, are not similar [16]. Besides, the second building resists the vibration induced in the first building, which may cause the collapse of both, in the case of miscalculations due to local failures under strong ground motions, particularly if one of the adjacent buildings is controlled using based-isolations [17].

Most of the recent studies have used simplified, one-dimensional shear frames or symmetric 3D buildings [18,19], neglecting the geometry and irregularity of structures, as well as bidirectional seismic loading. Earthquake loads can be applied in two directions that may cause simultaneous translational and torsional vibrations in buildings with considerable irregularity [20,21]. Such coupled vibrations can cause nonlinear behavior and severe damage in the external frame members, particularly, the corner columns and the bracing system [22,23]. Furthermore, for those structures built on a soft soil medium, different dynamic characteristics result in different responses due to the soil–structure interaction (SSI) [24–26]. Considering the SSI, Farshidifar and Soheili [27] studied the seismic response of actively controlled single and multi-story buildings; however, similar studies need to be carried out to consider the asymmetric properties of structures, as well as the pounding hazard of adjacent high-rise buildings. Therefore, despite the advantages of the proposed methods for regular buildings under the unidirectional seismic loads, such solutions are not always the best solution when two irregular buildings with different dynamic characteristics need to be controlled for possible pounding hazards.

As an alternative to a wired sensors network for smart structures, wireless sensor networks (WSNs) have attracted several researchers [28]. With lower cost, an array of WSNs can effectively be used for monitoring the structural deterioration, as well as controlling vibration during an earthquake, by using artificial neural networks [29,30]. The data that is acquired using WSNs is essentially the same as the traditional counterpart that has been challenged recently due to deployment difficulties, as well as reliability issues during an extreme event. Nowadays it is feasible to mimic the swarm behaviors in nature and share important response data with the adjacent buildings using a wireless sensors network and via cloud-based computing [31]. Keeping this in mind, the idea of swarm-based parallel control (SPC) of such buildings is proposed in this paper. With this approach, the response data from each wireless sensor is transferred to the adjacent buildings to update the responses without physical structural links. Figure 1 illustrates such information flow that is inspired by nature. Using the similar concept for adjacent buildings, the control force can be determined for each building with respect to the other surrounding structures, meaning that using the proposed algorithm, the optimal control forces are first determined based on the response of each building individually, and then, it is modified based on the responses of the other adjacent structures. In this study, the fuzzy inference system (FIS) is used to interpret the data using nonlinear mappings. The nature-inspired information flow among the swarms can be seen in Figure 1, while the information flow between each adjacent building is illustrated in Figure 2.

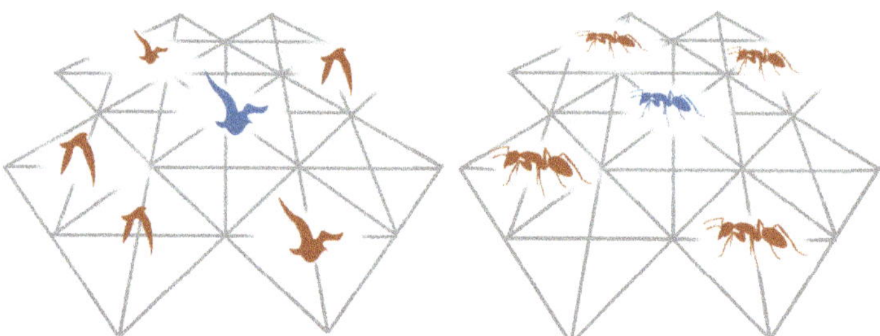

Figure 1. Examples of swarm intelligence in nature and information flow among individuals.

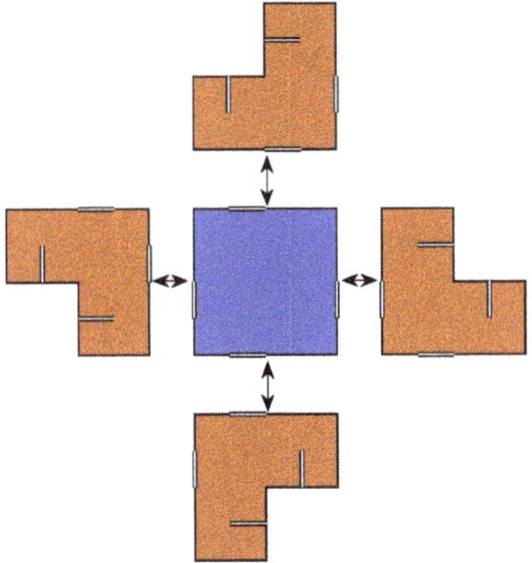

Figure 2. The information flow between five adjacent buildings for the proposed swarm-based parallel control (SPC) (plan view).

The goal of this study is to introduce and validate the performance of the proposed algorithm for adjacent buildings, using semi-active MR dampers that are wirelessly controlled by utilizing cloud-based computation. In addition, irregularity and soil–structure interactions are considered in this study. Two pairs of MR dampers are installed on each floor in two directions to suppress coupled translational–torsional motions, which are controlled by implementing the concept of the Internet-of-Things (IoT) and wireless sensor networks (WSNs). Robustness of the proposed method is evaluated and verified using three example cases: a single multi-story building considering the SSI, two adjacent buildings coupled using MR dampers as well as active actuators, and five adjacent buildings controlled using the proposed SPC. The LQR-based active control system is considered to provide benchmark results for comparison purposes only, and the mechanism and advantages of the proposed method over the active tendon controller are not discussed in this study.

2. Analytical Equations of Motion Considering Soil–structure Interaction

Figure 3 shows a multi-story shear frame building considering the soil–structure interaction simulation. In three-dimensional space, a building can be idealized as a (3 × n Story + 5)-DOF system. This is because each story has three degrees of freedom, plus five degrees of freedom for the base of the structure, swaying motion in the x- and y-directions, rocking about the x and y-axes, and twisting about the z-axis. As shown in Figure 3, pairs of dashpots and springs are defined to consider the SSI effects in the simulations. In this figure, x_b denotes the base displacement, ϕ_y represents the rocking motions about the y-axis, and the relative displacements of stories are shown as x_i. Note that the settlement in the z-direction is not considered, and the twisting motion is not visible from the side view.

The control devices placement is shown in Figure 4 for the MR devices, and due to the eccentricity with respect to the center of mass, d_i, each device generate a moment to resists the torsional motion. Irregularity in plan and elevation is described by the mass eccentricity with respect to the center of rigidity, e_x and e_y.

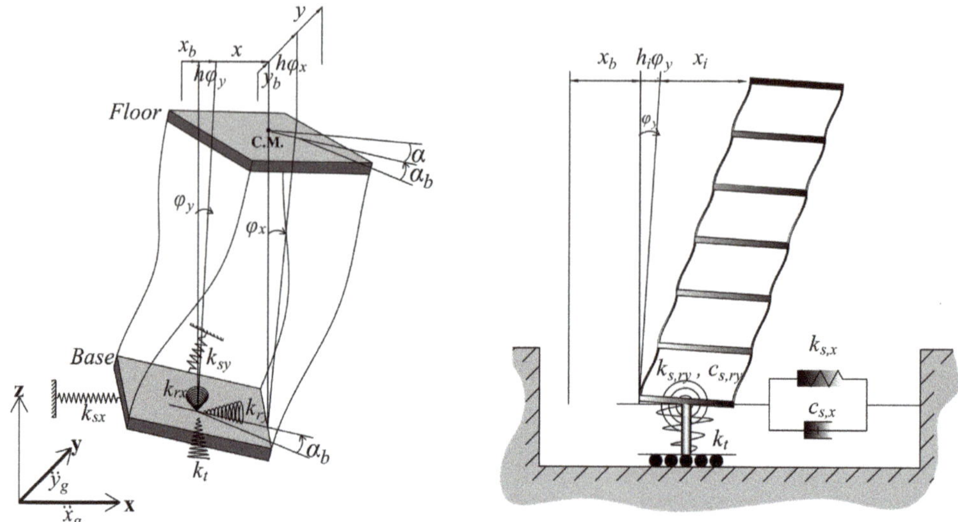

Figure 3. The idealized model of the soil–structure system.

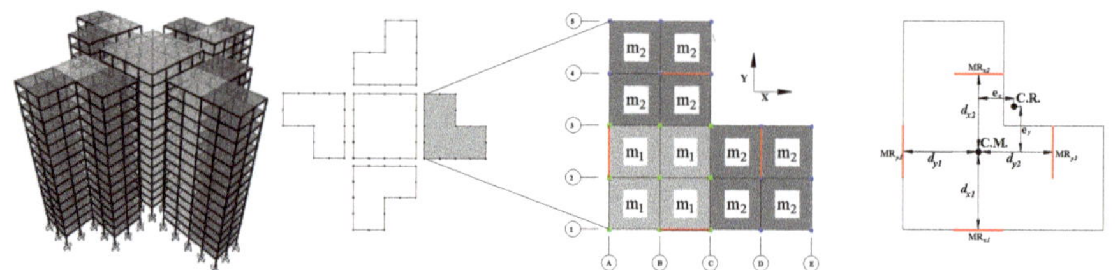

Figure 4. The configuration of the adjacent buildings used in this study.

The governing equations of motion for a controllable n-story building can be expressed as follows in general,

$$[M]\{\ddot{x}(t)\} + [C]\{\dot{x}(t)\} + [K]\{x(t)\} = [\gamma]\{u(t)\} + \{\delta\}\ddot{x}_g(t) \quad (1)$$

where $[M]$, $[C]$ and $[K]$ are the dynamic properties of the building, and $\{x(t)\}$, $\{\dot{x}(t)\}$ and $\{\ddot{x}(t)\}$ are the relative (to the ground motion) displacement, velocity and acceleration vectors, respectively. $[\gamma]$ is the coefficient matrix of the input controlling forces, and $\{\delta\}$ is the ground motion coefficient vector. The governing equation of motion can be re-written in state-space form as explained in [32]:

$$\{\dot{Z}(t)\} = [A]\{Z(t)\} + [B_u]\{u(t)\} + \{B_r\}\ddot{x}_g(t) \quad (2)$$

where

$$\{Z(t)\} = \left\{ \begin{array}{c} x(t) \\ \dot{x}(t) \end{array} \right\}, \quad (3a)$$

$$A = \left[\begin{array}{cc} [0] & I \\ -M^{-1}K & -M^{-1}C \end{array} \right], \quad (3b)$$

$$B_u = \left[\begin{array}{c} [0] \\ M^{-1}[\gamma] \end{array} \right], \quad (3c)$$

$$\{B_r\} = \left\{ \begin{array}{c} \{0\} \\ [M]^{-1}\{\delta\} \end{array} \right\}, \tag{3d}$$

It should be noted that the above-mentioned equations are in the continuous-time format; however, for the simulation with Matlab software, they are converted to the discrete-time format. A step-by-step procedure for computing the response of the system in the discrete-time domain is presented in the reference book by Cheng et al. [33]. The ground motions coefficients are given in $\{\delta\}$ for two directions, and the coefficient of the input controlling forces on each degree of freedom, including rotation, $\{u(t)\}$, can be described using the $[\gamma]$ matrix as [32,34]:

$$[\gamma] = \begin{bmatrix} [\gamma]_1 & -[\gamma]_2 & & & \\ & [\gamma]_2 & -[\gamma]_i & & \\ & & [\gamma]_i & \ddots & \\ & & & \ddots & -[\gamma]_n \\ & & & & [\gamma]_n \end{bmatrix}, \tag{4}$$

where $[\gamma]_i$ is

$$[\gamma]_i = \begin{bmatrix} 1 & 1 & 0 & 0 \\ 0 & 0 & 1 & 1 \\ d_{x1} & -d_{x2} & -d_{y1} & d_{y2} \end{bmatrix}. \tag{5}$$

Then, the mass and stiffness matrices considering the SSI effects can be assembled as:

$$[M] = \begin{bmatrix} [M]_{11} & [M]_{12} \\ [M]_{21} & [M]_{22} \end{bmatrix}_{(3n+5)\times(3n+5)} \tag{6}$$

where

$$[M]_{11} = \begin{bmatrix} [m]_1^* & [0] & [0] \\ [0] & \ddots & [0] \\ [0] & [0] & [m]_n^* \end{bmatrix}_{3n\times 3n}, [M]_{12} = [M]_{21}^T = \begin{bmatrix} m_1 & 0 & 0 & m_1h_1 & 0 \\ 0 & m_1 & 0 & 0 & m_1h_1 \\ 0 & 0 & I_{z,1} & 0 & 0 \\ \vdots & \vdots & \vdots & \vdots & \vdots \\ m_n & 0 & 0 & m_nh_n & 0 \\ 0 & m_n & 0 & 0 & m_nh_n \\ 0 & 0 & I_{z,n} & 0 & 0 \end{bmatrix}_{3n\times 5}$$

$$M_{22} = \begin{bmatrix} \left(\sum\limits_{i=1}^{n} m_i\right) + m_b & 0 & 0 & \sum\limits_{i=1}^{n} m_ih_i & 0 \\ 0 & \left(\sum\limits_{i=1}^{n} m_i\right) + m_b & 0 & 0 & \sum\limits_{i=1}^{n} m_ih_i \\ 0 & 0 & \sum\limits_{i=1}^{n} I_{z,i} + I_{z,b} & 0 & 0 \\ \sum\limits_{i=1}^{n} m_ih_i & 0 & 0 & \sum\limits_{i=1}^{n} \left(I_{y,i} + m_ih_i^2\right) + I_{y,b} & 0 \\ 0 & \sum\limits_{i=1}^{n} m_ih_i & 0 & 0 & \sum\limits_{i=1}^{n} \left(I_{x,i} + m_ih_i^2\right) + I_{x,b} \end{bmatrix}_{5\times 5} \tag{7}$$

where $[0]$ is a zero matrix, and $[m]_i^*$ is the mass matrix of the ith story [32].

$$[m]_i^* = \begin{bmatrix} m_i & 0 & 0 \\ 0 & m_i & 0 \\ 0 & 0 & I_{z,i} \end{bmatrix} \tag{8}$$

In the above-mentioned equations, m_i is the mass of the ith story, the mass moment inertia about each axis in three-dimension is defined as follows for a floor slab with dimensions of $a \times b$, and coordinates of $\left(x_{m,j}, y_{m,j}\right)$. The center of the mass coordinate is given as $(\overline{X}_m, \overline{Y}_m)$.

$$I_x = \sum_{j=1}^{jm}\left[\frac{m_j}{12}\left(b^2\right) + m_j\left[\left(x_{m,j} - \overline{X}_m\right)^2 + \left(y_{m,j} - \overline{Y}_m\right)^2\right]\right] \tag{9a}$$

$$I_y = \sum_{j=1}^{jm}\left[\frac{m_j}{12}\left(a^2\right) + m_j\left[\left(x_{m,j} - \overline{X}_m\right)^2 + \left(y_{m,j} - \overline{Y}_m\right)^2\right]\right] \tag{9b}$$

$$I_z = \sum_{j=1}^{jm}\left[\frac{m_j}{12}\left(a^2 + b^2\right) + m_j\left[\left(x_{m,j} - \overline{X}_m\right)^2 + \left(y_{m,j} - \overline{Y}_m\right)^2\right]\right] \tag{9c}$$

The stiffness matrix of each floor is not diagonal like the mass matrix [32].

$$[k]_i^* = \begin{bmatrix} k_{xx} & 0 & k_{x\theta} \\ 0 & k_{yy} & k_{y\theta} \\ k_{\theta x} & k_{\theta y} & k_{\theta\theta} \end{bmatrix} \tag{10}$$

where

$$k_{xx} = \sum_{j=1}^{nk} k_{x,j}, \quad k_{yy} = \sum_{j=1}^{nk} k_{y,j} \tag{11a}$$

$$k_{x\theta} = k_{\theta x} = \sum_{j=1}^{nk} k_{x,j}\left(\overline{Y}_k - y_{k,j}\right) \tag{11b}$$

$$k_{y\theta} = k_{\theta y} = \sum_{j=1}^{nk} k_{y,j}\left(\overline{X}_k - x_{k,j}\right) \tag{11c}$$

$$k_{\theta\theta} = \sum_{j=1}^{jk}\left(k_{x,j}\left(\overline{Y} - y_{k,j}\right)^2 + k_{y,j}\left(\overline{X} - x_{k,j}\right)^2\right), \tag{11d}$$

Similarly, the center of stiffness is located at $(\overline{X}_k, \overline{Y}_k)$, and that of the jth lateral resisting member with the coordinate $\left(x_{k,j}, y_{k,j}\right)$ has the stiffness of $k_{x,j}$, and $k_{y,j}$ in two directions. The jth column is assumed to have $k_{x,j} = k_{y,j} = 12EI_j/h^3$. Thus, the stiffness matrix is assembled as follows [32,34]:

$$[K] = \begin{bmatrix} [K]_{\text{superstructure}} & [0] \\ [0]^T & [K]_{\text{soil}} \end{bmatrix} \tag{12a}$$

$$[K]_{\text{superstructure}} = \begin{bmatrix} [k]_1^* + [k]_2^* & -[k]_2^* & [0] & [0] & [0] \\ -[k]_2^* & [k]_2^* + [k]_3^* & -[k]_3^* & [0] & [0] \\ [0] & -[k]_3^* & \ddots & \vdots & \vdots \\ [0] & [0] & \cdots & \ddots & -[k]_n^* \\ [0] & [0] & \cdots & -[k]_n^* & [k]_n^* \end{bmatrix} \tag{12b}$$

$$[K]_{\text{soil}} = \text{diag}\left(k_{x,s}, k_{y,s}, k_{t,s}, k_{rx,s}, k_{ry,s}\right) \tag{12c}$$

The damping matrix is constructed in the same way as for the stiffness matrix, but the superstructure damping is determined using Rayleigh's method [35] without SSI. The SSI parameters are determined using the equations given in [25,32,34], which are summarized in Table 1. In this study, two cases with and without SSI effects are investigated. The soft soil is assumed to have the Poisson's ratio of 0.33, the density of 2400 kg/m^2, shear-wave velocity and modulus of 500 m/s and 6 × 108 N/m^2, respectively. Details of the SSI models, as well as the methodology, are presented by Nazarimofrad and Zahrai [34].

Table 1. Dynamic properties of the springs and dashpots in the swaying, rocking and twisting directions for simulating soil–structure interaction (SSI) effects [32].

Motion	r	Stiffness	Damping
Swaying	$\sqrt{\frac{A_0}{\pi}}$	$k_s = \frac{8\rho V_s^2 r}{2-\nu}$	$c_s = \frac{4.4r^2}{2-\nu}.\rho V_s$
Rocking	$\sqrt[4]{\frac{4I_{x(\text{or } y)}}{\pi}}$	$k_r = \frac{8\rho V_s^2 r^3}{3(1-\nu)}$	$c_r = \frac{0.4r^4}{1-\nu}.\rho V_s$
Twisting	$\sqrt[4]{\frac{2I_z}{\pi}}$	$k_t = \frac{16\rho V_s^2 r^3}{3}$	$c_t = 0.8r^4 \rho V_s$

3. Swarm-Based Parallel Control (SPC)

In this study, three different control alternatives for adjacent buildings are presented and discussed. One most commonly used method is to control each building independently without any information flow (Case-I in this paper). Another approach is to couple two adjacent buildings side-by-side by actuators or additional dampers, such as hydraulic or MR dampers (Case-II in this paper). Both methods are discussed, and the results are compared with the proposed swarm-based parallel control (SPC), which is Case-III in this paper. The SPC algorithm is the FLC for each building, with an additional updating agent that is another FLC to consider the response of the other buildings simultaneously.

For Case-I, the behavior and control of a single building (in the east, Figure 4) are studied considering the soil–structure interaction (SSI) effects, and the performance of the different control algorithms are compared when the behavior of the adjacent structure is neglected. The response reduction ratio is considered as the performance index, and since the active control has a different control mechanism, the ideal condition is considered for this study. The linear quadratic regulator (LQR) algorithm is used for calculating the optimal control forces that can be applied to the structure through pairs of active tendons that can be installed at the same locations as for the MR dampers. Therefore, the advantages of FLC and SPC are not compared to the active control system. The fuzzy logic control (FLC) is used to efficiently control vibrations using semi-active MR dampers, which is illustrated in Figure 5. More details, and the background of LQR and FLC algorithms are available from refs. [32,36,37]. The weight matrices for the LQR method are selected as $R = 10^{-6} \times I_{\text{size}([\gamma])}$, and $Q = 10^6 \times I_{2n}$, where I is an identity matrix.

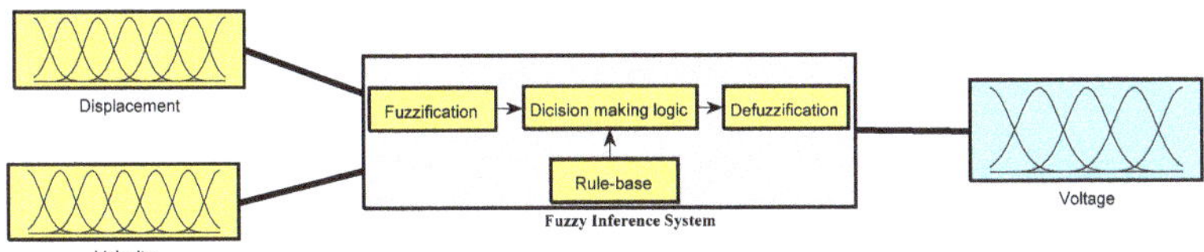

Figure 5. The fuzzy logic control block diagram (Type: Mamdani).

Figure 4 shows five adjacent buildings with different dynamic properties. Instead of coupling buildings (Case-II), response data is shared among the controllers. Thus, an additional control fuzzy logic control layer is considered to avoid pounding, for which the input variables are the relative displacements and the directions. Using this strategy, the optimal required clearance distance can be decreased, which is critical in urban areas, where the cost of space and material is essential. Using this method, the adjacent buildings are not connected physically, and the response data are shared wirelessly. Thus, buildings are independently controlled, and then their relative performances are evaluated using a second fuzzy logic control algorithm. In this paper, the first fuzzy logic control, FIS1, determines the control forces for individual buildings independently, and the second fuzzy logic control, FIS2, determines the control forces based on the relative motions including the torsion in each irregular building. Finally, the optimal MR dampers input voltages are selected by comparing the outputs of FIS1 and FIS2.

The two surface plots of the outputs are shown in Figure 6 for FIS1 and FIS2 using the Gaussian membership functions for the variables. In Figure 7, the flowchart of the proposed swarm-based control (SPC) is described in summary. The flowchart includes the following steps:

Step 1. Estimating the dynamic properties of the soil as well as the building (e.g., using data-driven-based machine learning models [38]).

Step 2. Assembling the state-space model and equation of motions according to Equations (1)–(12).

Step 3. Calculating the response of each of the adjacent buildings, $\{x(t)\}$, by solving the state-space equations in Step 3.

Step 4. Normalizing the state vector (i.e., measures responses, $\{x(t)\}$ and calculating the control force using the fuzzy logic control FIS1.

Step 5. Updating the calculated control force in both directions, using the second fuzzy logic-based updating rules, FIS2, based on the relative displacements of two adjacent buildings. In this step, the outputs of FIS1 and FIS2 are compared, and if the FIS2 output is higher than the FIS1 output, the control force is adjusted based on the FIS2 results.

Step 6. Apply the calculated control force, $\{u(t)\}$, and continuing until the end of the last time-step.

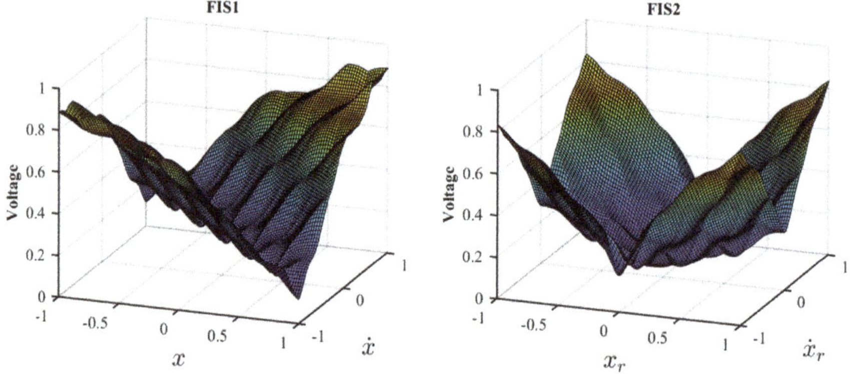

Figure 6. The surface plot of the FIS1 and FIS2, which are used in the SPC algorithm.

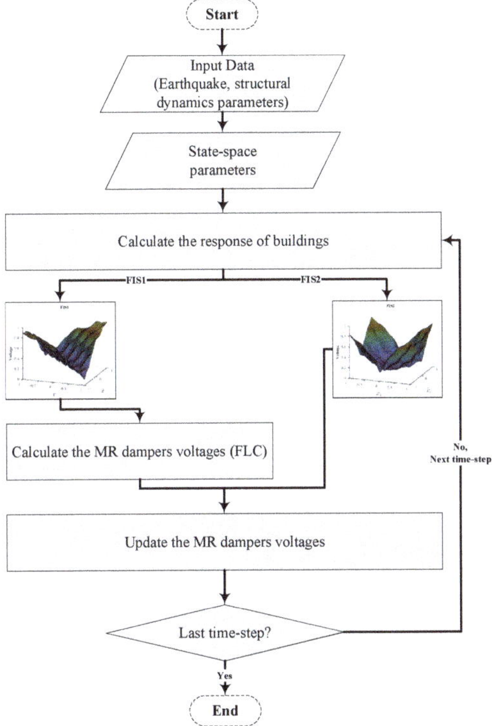

Figure 7. The flowchart of the proposed swarm-based parallel control (SPC).

4. Numerical Studies

4.1. Building Models Discerptions

In this paper, four adjacent 15-story irregular steel frame buildings were designed and simulated in MATLAB® (MathWorks, Inc., Nattick, MA, USA), based on the studies carried out by Kaveh et al. [39] and Azimi [32]. The 3D and plan views of the structures, as well as the mass and stiffness distributions, are visually illustrated in Figure 4. Two groups of columns, with the same stiffness in both directions, are connected rigidly to the floor diaphragm frames, $k_1 = 2k_2 = 1200$ kN/m . Each floor consists of 5×5 m^2 slabs that carry loads of $m_1 g = 2m_2 g = 30$ (kN/m^2) to simulate the eccentricity. Each story has a height of 3.0 m, except the first story, with a 3.2 m height. The four dampers of each floor are placed within the frames shown with red bold lines. To ensure the accuracy of the simulation results, the building model simulation is verified, based on the study by Nazarimofrad et al. [34]. Figure 8 shows the top floor level response of the 10-story building model under the same ground motion, which indicates that the simulation results in this study are valid.

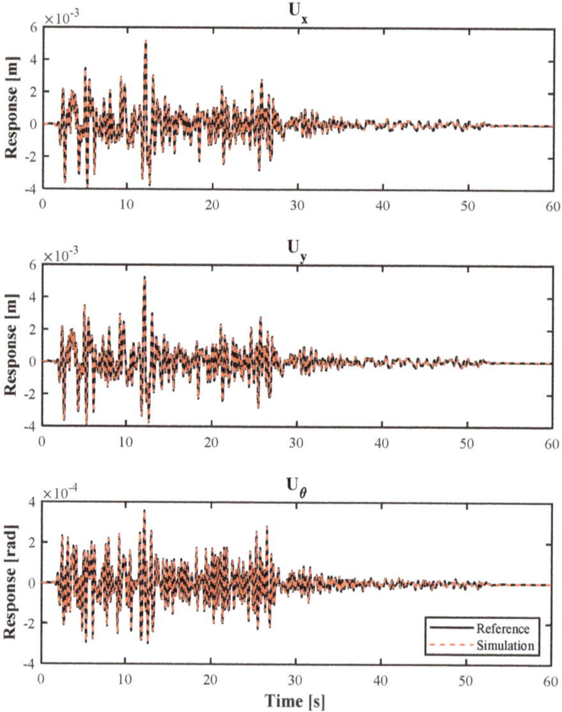

Figure 8. Verification of the simulation for an uncontrolled 10-story building based on Nazarimofrad and Zahrai [34].

4.2. Magneto-Rheological (MR) Damper

The phenomenological model for the MR damper was first proposed by Spencer et al. [40], based on the Bouc–Wen model in 1997. Nowadays, the MR damper is one of the most reliable devices for seismic vibration control applications. The advantage of the MR damper is its role in the case of power loss, which provides minimum damping based on its shaft displacement and velocity. Using a semi-active control algorithm, the control command is the input voltage of the MR damper. The modified model of MR damper is given in Figure 9. The notation x in Figure 9 and Equations (13)–(16) denotes the relative displacement of the two ends of each MR damper.

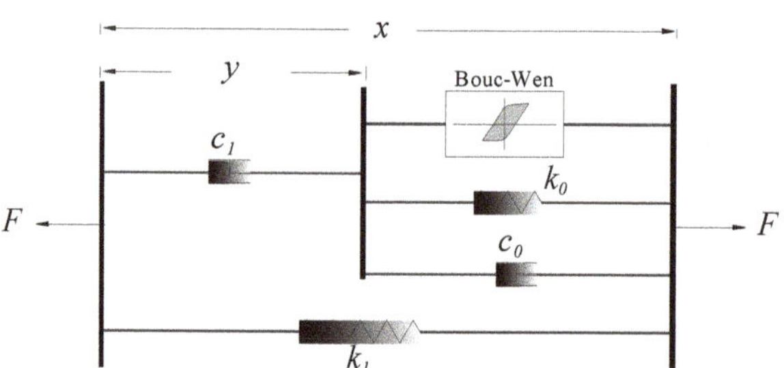

Figure 9. Modified Bouc–Wen model of the magnetorheological (MR) damper (replotted based on Nugroho et al. [41]).

More details are available in the literature regarding the history of the development of the analytical model of MR dampers. According to Spencer et al. [40], the nonlinear behavior of an MR damper can be described as follows

$$F = c_1\dot{y} + k_1(x - x_0) \tag{13}$$

$$\dot{y} = \frac{1}{c_0 + c_1} + \left(\alpha z + c_0\dot{x} + k_0(x - y)\right) \tag{14}$$

where F is the nonlinear force of the MR damper (included in $\{u(t)\}$), k_1 is the accumulator stiffness, and z is the evolutionary variable of the hysteretic, defined as:

$$\dot{z} = -\gamma\left|\dot{x} - \dot{y}\right|z|z|^{n-1} - \beta\left(\dot{x} - \dot{y}\right)|z|^n + A\left(\dot{x} - \dot{y}\right), \tag{15}$$

where γ, β and A are the shape parameters, and the α, c_0 and c_1 can be obtained using the following linear functions of the efficient voltage, u.

$$\alpha(u) = \alpha_a + \alpha_b u \tag{16a}$$

$$c_0(u) = c_{0a} + c_{0b}u \tag{16b}$$

$$c_1(u) = c_{1a} + c_{1b}u \tag{16c}$$

$$\dot{u} = -\eta(u - v) \tag{16d}$$

The parameters of the MR damper are given in Table 2.

Table 2. MR damper parameters [32,40].

Parameter	Value	Parameter	Value
$c0_a$	50,300 (N s/m)	α_a	8700 (N/m)
$c0_b$	48,700 (N s/m V)	α_b	6400 (N/m V)
$c1_a$	8,106,200 (N s/m)	γ	496 m^{-2}
$c1_b$	7,807,900 (N s/m V)	β	496 m^{-2}
k_0	5.4E-6 (N/m)	A	810.50
k_1	8.7E-6 (N/m)	n	2
x_0	0.18 (m)	η	190 s^{-1}

4.3. Earthquake Loads

In this study, seven earthquake records have been selected to compare the performance of each control method. The characteristic properties of the records are given in Table 3. The elastic acceleration response spectra are shown in Figure 10, which is also used in the previous studies [32,36,37].

Table 3. Characteristics of the selected earthquake records.

Earthquake [1]	Station and Direction	Magnitude (Mw)	PGA (g)	PGV (cm/s)
1940 El Centro	El Centro Array #9 270°	7.2	0.21	31.3
	El Centro Array #9 180°	7.2	0.28	30.9
1994 Northridge	Sylmar—Olive View Med FF 360°	6.7	0.84	129.6
	Sylmar—Olive View Med FF 090°	6.7	0.61	77.5
1995 Kobe	H1170546.KOB 090°	7.2	0.63	76.1
	H1170546.KOB 000°	7.2	0.83	91.1
1999 Chi-Chi	TCU068 N	7.6	0.37	264.1
	TCU068 E	7.6	0.51	249.6
1971 San Fernando	Pacoima Dam 164°	6.6	1.22	114.5
	Pacoima Dam 254°	6.6	1.24	57.3

Table 3. *Cont.*

Earthquake [1]	Station and Direction	Magnitude (Mw)	PGA (g)	PGV (cm/s)
1989 Loma Prieta	Hollister—South and Pine 0°	6.9	0.37	63.0
	Hollister—South and Pine 0°	6.9	0.18	30.9
1992 Erzincan	Erzincan—EW	6.7	0.50	78.2
	Erzincan—NS	6.7	0.39	107.14

[1] Source: http://ngawest2.berkeley.edu/.

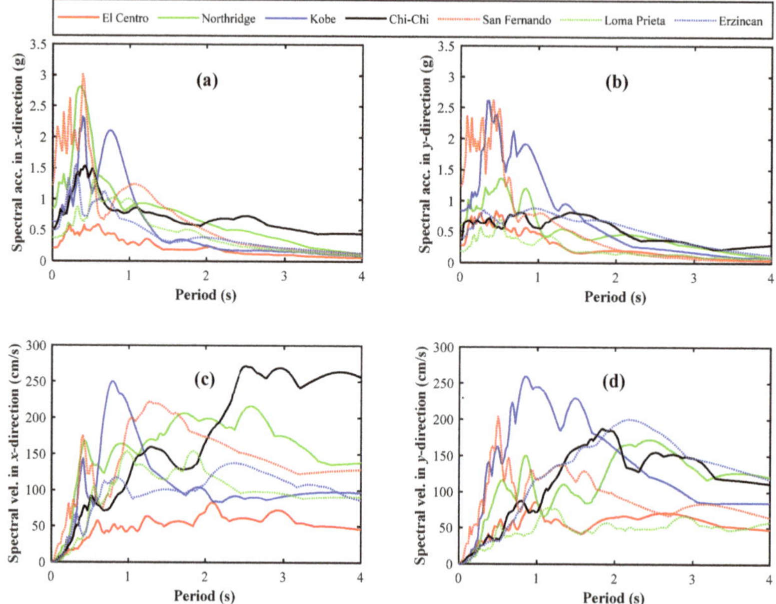

Figure 10. Spectral acceleration, (**a**) and (**b**), and spectral velocity, (**c**) and (**d**), of the selected earthquakes in two directions.

5. Results and Discussion

Irregularity in high-rise buildings is an important factor in evaluating the seismic responses, due to the higher modes effects under bidirectional seismic loads [32]. Under the bidirectional earthquake loads, it is clear that higher modes of an irregular structure will have a significant impact on the results, particularly considering the soil–structure interaction effect. In the following sections, the structural responses of the buildings are discussed in detail for the three chases.

5.1. Control of Single Building Considering Soil–structure Interaction (Case I)

5.1.1. Peak Responses

The peak responses of the Case-I building under the seven selected earthquake records are given in Table 4, which includes the maximum top floor transversal and rotational displacements, and also the accelerations in two directions. The optimal active control responses are obtained using a full-state LQR controller. For the passive-min and passive-max cases, also known as passive-off and passive-on, the input voltage for each MR damper is constant and equal to 0 and 9 volts, respectively. In the case of power loss, and under the seven earthquakes given in the table, the passive-min case offers 13%, 8%, 10%, 3%, 13%, 8% and 14% reduction in transversal displacements, when the building is on soft soil. The response reduction percentages are approximately 30% smaller for the same building on the rock base.

Table 4. Maximum responses of the Case-I building using different control methods considering SSI effects (units in meters).

Earthquake	Max. Response [1]	Uncontrolled		Active		Passive-min		Passive-max		FLC	
El Centro	Displacement$_x$	0.15	(0.31)	0.08	(0.11)	0.13	(0.28)	0.08	(0.20)	0.07	(0.17)
	Displacement$_y$	0.14	(0.28)	0.05	(0.07)	0.13	(0.26)	0.06	(0.14)	0.06	(0.13)
	Rotation	0.006	(0.011)	0.004	(0.004)	0.005	(0.010)	0.002	(0.005)	0.002	(0.005)
	Acceleration$_x$	1.52	(2.72)	0.54	(1.94)	1.16	(2.58)	1.39	(2.41)	1.72	(2.78)
	Acceleration$_y$	2.18	(3.49)	0.89	(2.96)	2.01	(3.41)	1.59	(3.30)	1.83	(3.50)
Northridge	Displacement$_x$	0.42	(0.75)	0.20	(0.28)	0.40	(0.73)	0.27	(0.60)	0.25	(0.55)
	Displacement$_y$	0.26	(0.48)	0.10	(0.22)	0.24	(0.48)	0.15	(0.41)	0.16	(0.40)
	Rotation	0.037	(0.074)	0.014	(0.026)	0.035	(0.071)	0.020	(0.053)	0.020	(0.053)
	Acceleration$_x$	5.37	(11.71)	2.19	(8.41)	5.07	(11.54)	4.29	(10.97)	3.60	(10.19)
	Acceleration$_y$	3.77	(7.21)	1.37	(6.17)	3.15	(7.00)	3.12	(7.15)	3.01	(6.92)
Kobe	Displacement$_x$	0.29	(0.48)	0.08	(0.14)	0.26	(0.45)	0.13	(0.28)	0.13	(0.28)
	Displacement$_y$	0.27	(0.46)	0.10	(0.17)	0.25	(0.43)	0.14	(0.34)	0.14	(0.34)
	Rotation	0.008	(0.014)	0.003	(0.004)	0.007	(0.013)	0.003	(0.008)	0.004	(0.009)
	Acceleration$_x$	6.24	(7.41)	1.18	(5.63)	5.33	(6.45)	2.95	(7.05)	3.29	(7.40)
	Acceleration$_y$	5.54	(9.54)	1.80	(8.09)	4.78	(9.08)	3.14	(8.93)	3.12	(8.69)
Chi-Chi	Displacement$_x$	0.66	(1.56)	0.87	(0.53)	0.64	(1.53)	0.48	(1.27)	0.41	(1.11)
	Displacement$_y$	0.39	(0.91)	0.99	(0.47)	0.38	(0.91)	0.33	(0.83)	0.29	(0.73)
	Rotation	0.042	(0.089)	0.030	(0.025)	0.040	(0.086)	0.025	(0.066)	0.027	(0.064)
	Acceleration$_x$	4.66	(7.85)	1.33	(5.88)	4.00	(7.78)	2.24	(7.02)	2.35	(6.94)
	Acceleration$_y$	3.22	(6.62)	1.44	(5.49)	2.85	(6.48)	2.47	(6.20)	2.53	(5.78)
San Fernando	Displacement$_x$	0.33	(0.62)	0.16	(0.31)	0.32	(0.60)	0.24	(0.52)	0.22	(0.47)
	Displacement$_y$	0.15	(0.19)	0.05	(0.12)	0.13	(0.17)	0.08	(0.17)	0.08	(0.17)
	Rotation	0.017	(0.030)	0.009	(0.015)	0.016	(0.029)	0.010	(0.021)	0.010	(0.020)
	Acceleration$_x$	5.28	(11.44)	2.65	(11.79)	4.83	(11.30)	4.05	(10.92)	3.76	(11.06)
	Acceleration$_y$	3.95	(12.12)	2.87	(11.94)	3.38	(12.40)	3.56	(12.17)	3.55	(12.91)
Loma Prieta	Displacement$_x$	0.29	(0.52)	0.13	(0.22)	0.28	(0.51)	0.20	(0.43)	0.18	(0.39)
	Displacement$_y$	0.13	(0.32)	0.11	(0.10)	0.12	(0.28)	0.08	(0.17)	0.08	(0.15)
	Rotation	0.018	(0.035)	0.009	(0.014)	0.015	(0.032)	0.009	(0.020)	0.008	(0.018)
	Acceleration$_x$	3.33	(6.25)	1.12	(4.00)	3.09	(5.99)	2.46	(5.58)	2.15	(5.16)
	Acceleration$_y$	1.50	(1.96)	0.56	(1.71)	1.13	(2.15)	1.29	(2.17)	1.44	(2.27)
Erzincan	Displacement$_x$	0.29	(0.63)	0.13	(0.20)	0.25	(0.58)	0.16	(0.44)	0.14	(0.36)
	Displacement$_y$	0.41	(0.77)	0.19	(0.28)	0.39	(0.76)	0.27	(0.64)	0.25	(0.57)
	Rotation	0.030	(0.066)	0.016	(0.023)	0.029	(0.064)	0.019	(0.049)	0.018	(0.045)
	Acceleration$_x$	2.94	(6.21)	1.14	(4.88)	2.21	(5.93)	1.94	(4.92)	1.84	(4.42)
	Acceleration$_y$	4.00	(6.00)	1.57	(5.20)	3.89	(6.05)	2.77	(6.08)	2.91	(5.76)

[1] Values in parenthesis are the corresponding responses for the rock base (no SSI).

The passive-max case, however, provides 40–50% more reduced compared with the passive-min control. Using the FLC offers even more maximum response reduction, which is about 60%, 40%, 55%, 38%, 47%, 38% and 52% for the building on soft soil under the seven selected earthquakes, respectively. For the building on a rock base, the FLC is 10% less efficient, however. Comparing the maximum responses for the active control scenario, where the controlling tendons are assumed to deliver the real-time control forces with minimum errors, reveals that this method does not provide the same performance under the Chi-Chi earthquake for the same building on rock and soft soil bases. Despite the 50–66% displacement response reduction for the building on the rock base, active control causes the displacement responses to be increased by 32% and 154% in two directions. In general, it can be seen that the peak responses are smaller for the buildings on the soft soil, except for the active control under the Chi-Chi earthquake. Therefore, the FLC offers reliable performance from this point of view, and it is used for the Parallel control of adjacent buildings.

5.1.2. Time-History Responses

The time-history responses provide a tool to understand the performance of any techniques during a seismic event [42]. Figure 11 shows the top floor displacement time-history in the x-, y-, and θ-directions for the Case-I building considering SSI under the Chi-Chi earthquake. It is clear that the designed active control for a building on a rock base does not necessarily have similar performance for the same building, but on a soft soil base. This different behavior shows the sensitivity of the active control system, as well as the reliability of the FLC method. For all the seven earthquakes, the FLC provided a better performance with and without considering SSI. Another important fact, evident from the peak responses table and the time-history curves, is that for those buildings on a soft soil base, the peak responses are smaller. By implementing a properly designed FLC, the overall response can be reduced by 40% and 50% for near- and far-field earthquakes, respectively [32].

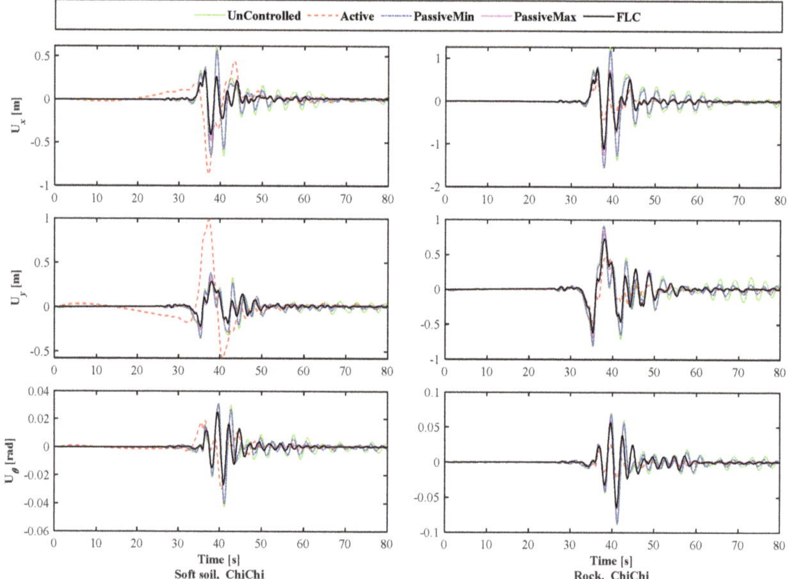

Figure 11. The time-history responses of the Case-I building considering SSI effects under the Chi-Chi earthquake.

5.1.3. Inter-Story Drift and Lateral Displacement Profiles

The inter-story drift profiles are typically used as a measure for the optimal placement of control devices, as well as estimating the potential damage locations. Smaller inter-story drifts, on the other hand, indicate that linear control algorithms can be satisfactory. Furthermore, P-Δ effects are reduced by decreasing the inter-story in the lower levels of a multi-story building. Smaller inter-story values for top floors makes it reasonable to use twin tuned mass dampers (TTMDs) instead of semi-active dampers, since the performance of semi-active devices is directly related to the relative displacement and velocity [37]. Figures 12 and 13 show the maximum inter-story drift profiles for the Case-I building in x-direction, under the bidirectional earthquakes loads. For this specific example, the active control using LQR algorithms shows a better performance in most of the cases.

The Chi-Chi earthquake is a good example to explain two different behaviors for the same controller, which highlights the importance of the additional five degrees of freedom (DOF) from the SSI. Similar differences can be seen under the El-Centro earthquake. From the displacement profiles, it is evident that in the case of power loss, Passive-Min, MR dampers still contribute to the response reduction by providing a small amount of damping [36]. The absolute displacement profiles provide useful tools to better understand the vibration control of adjacent buildings that are discussed in Case-II and Case-III buildings.

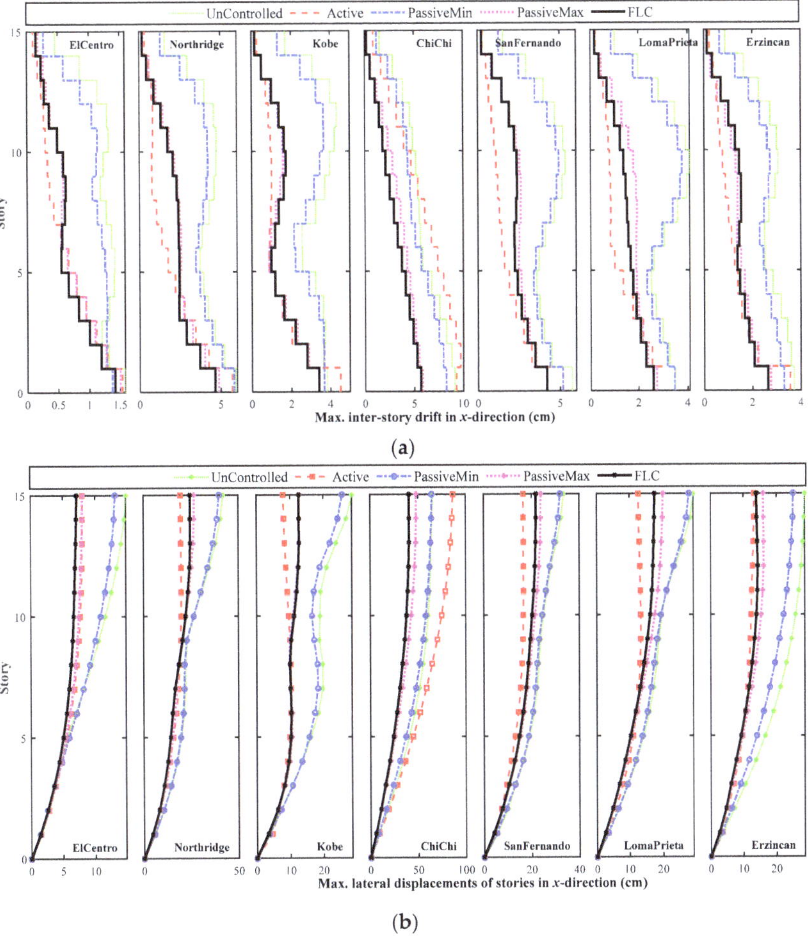

Figure 12. Maximum lateral inter-story drifts, (**a**), and maximum lateral displacements, (**b**), of the Case-I building in the x-direction (soft soil).

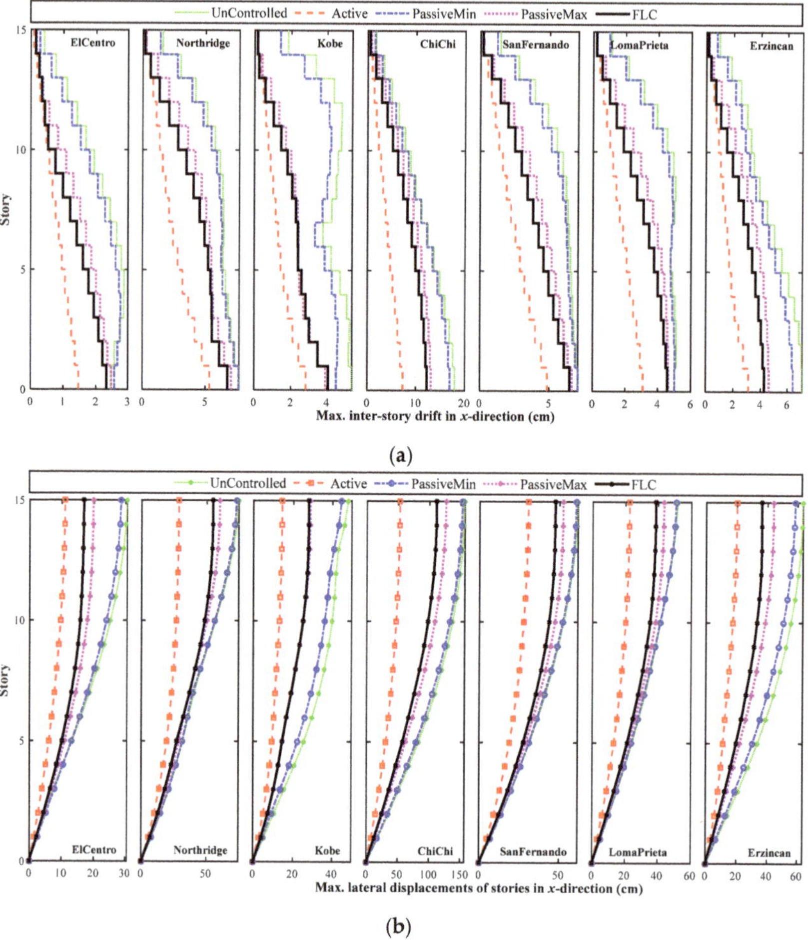

Figure 13. Maximum lateral inter-story drifts, (**a**), and maximum lateral displacements, (**b**), of the Case-I building in the x-direction (rock base).

Estimating the soil properties, as well as the dynamic characteristics of buildings, are not always accurate. To study the sensitivity of the structural responses, as well as the performance of the controller under uncertainties, a total of 120 random building models were generated, using 14 random variables for soil and structure models. The distribution of each random variable is assumed to be a Normal Gaussian with a coefficient of variation of 5%. In addition, 1% noise added to all the measurements. The results are provided in Figure 14, which includes the maximum lateral response profiles from each analysis, as well as the response reduction distribution for the top floor level in both directions. According to the results, it can be said that the FLC is less sensitive to the soil type based on the range of the histogram plots for the response reduction percentages for each case, however, the active control is highly sensitive to the estimated dynamic properties of the system. Therefore, fuzzy logic-based controllers are suitable for the proposed approach in this study. It is worth it to mention that the response variation in y-direction for the FLC controlled is higher than the x-direction, however, both have similar distribution as for the uncontrolled cases.

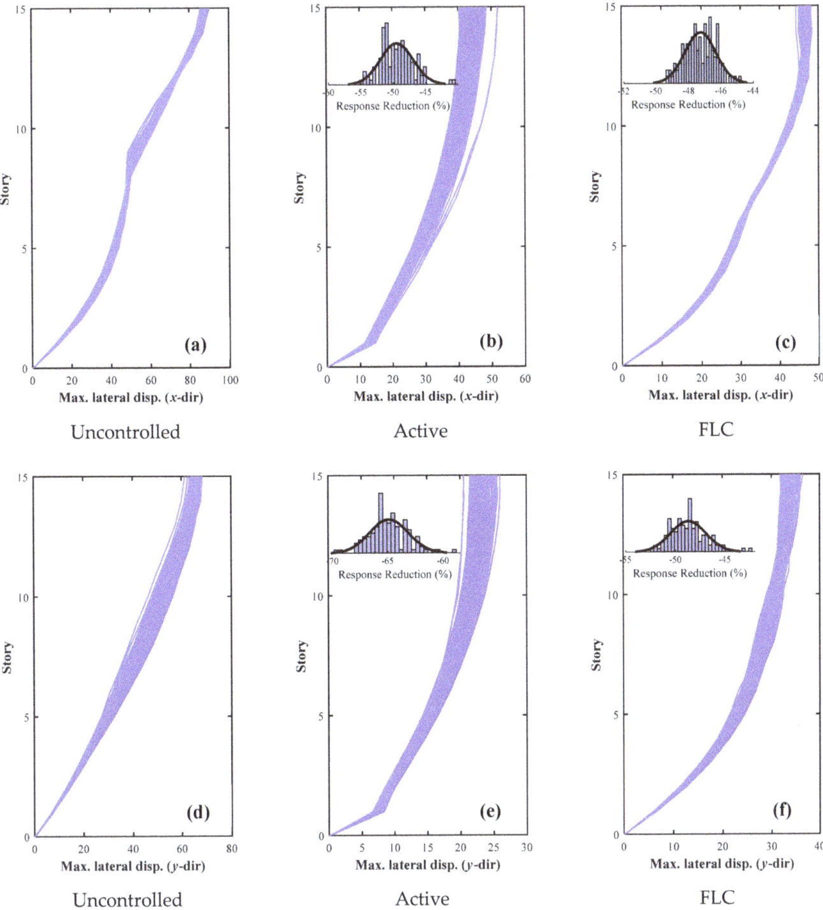

Figure 14. Maximum lateral displacements of the Case-I building considering uncertainties in dynamic properties (soft soil). (**a**) and (**d**): Uncontrolled; (**b**) and (**e**): Active LQR; (**c**) and (**f**): FLC.

5.2. Advantages of WSNs in Case of Structural Damage or Control Failure

One of the main drawbacks of the traditional wired sensor networks, is that they are not always reliable in case of fire, or due to the failure of signal lines at lower floor levels under large deformations. On the other hand, IoT and cloud-based computation, along with wireless sensor networks, make each device independently controllable and more reliable. The effect of the failure in the signal line in the traditional wired and wireless network are compared in Figure 15. In the wired sensor network, expecting that the main controlling computer remains safe and secured during an earthquake, the signal line is assumed to be broken at either the 4th or 6th floor, which would put the upper floors controlling devices in passive-min mode. However, using the IoT concept with a wireless sensors network, only the specified floor dampers are removed from the control algorithm. The overall structural responses of the building on the top floor in x-, y-, and θ-directions using the FLC indicate the superior performance of a wireless sensor network as part of the IoT. Comparing the responses also reveals that placing MR dampers at the first lower floors offers more resistance and damping during an earthquake, which is reasonable referring to the inter-story drift profiles.

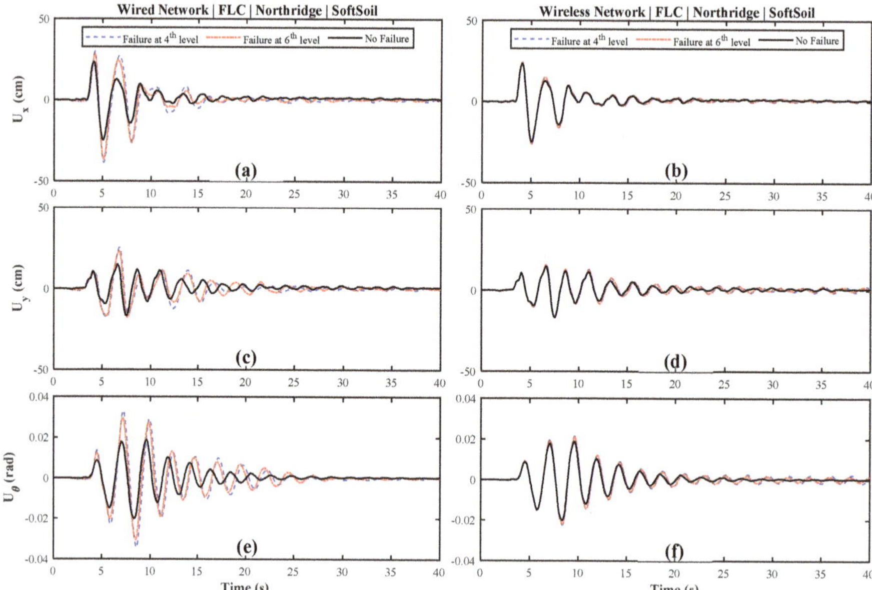

Figure 15. Wireless vs. wired sensor network in case of power or signal line failure (Case I) in the x-direction, (**a**) and (**b**), in the y-direction, (**c**) and (**d**), and in the θ-direction, (**e**) and (**f**).

5.3. Pounding Hazard Mitigation by Coupling (Case-II)

Design of a vibration control algorithm, when at least one of the two adjacent buildings is irregular, is a challenging task that needs knowledge and experience. To reveal one of the main challenges of designing an active control using active tendons to couple two buildings, which includes the optimal selection of the weights, the vibration responses of the two adjacent buildings, in the east (irregular) and the center (regular), are given in Figure 16. As it is clear from the figure, under the Northridge earthquake loads, the vibration of both buildings decreases significantly, as well as a considerable decrease in the minimum clear distance between the two buildings, considering the irregularity. These results are obtained when the parameters of the LQR algorithm is selected by an optimization technique. With this technique, the irregular building in the east passes a portion of lateral loads to the regular building to reduce the overall response of both, which can be seen from the results that indicate a minimum reduction in the peak displacement in the directions to the adjacent structure.

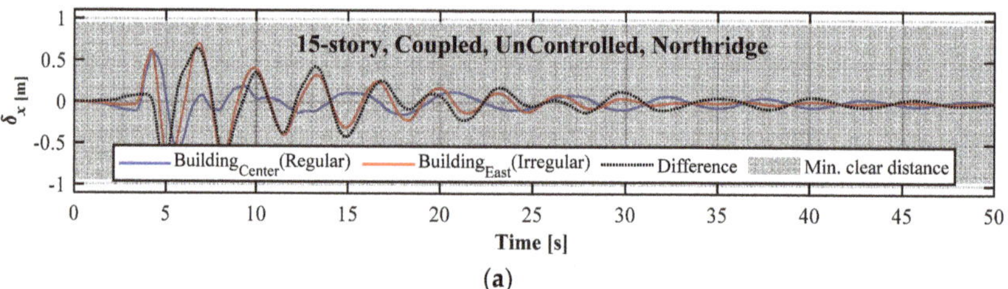

Figure 16. *Cont.*

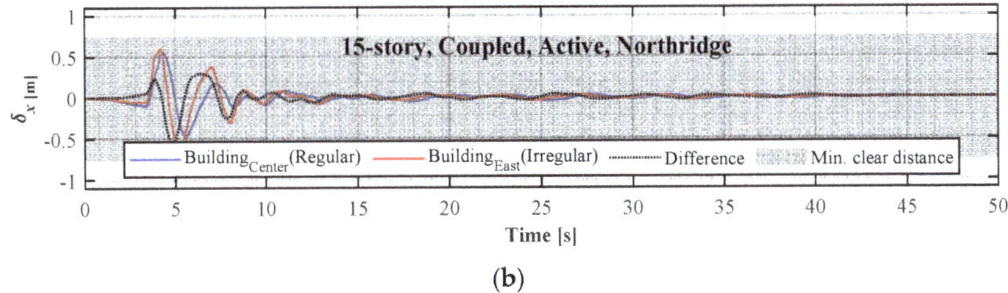

(**b**)

Figure 16. Top floor displacements of two coupled buildings on a rock base with uncontrolled, (**a**), active control algorithm, (**b**).

Figure 17, on the other hand, clearly shows that using an LQR-based active control for two coupled regular-irregular buildings are sensitive to the initial assessment of the dynamic properties of the system, as well as the selection of the LQR parameters. In this figure, two curves illustrate how the minimum clear distance between the two buildings increases as the weight coefficient on the response of specific DOFs in a direction changes, without updating the response weights in the other two directions. In this example, the response weight corresponding to the DOFs in the x-direction, as well as another direction (y or θ), is given a unit coefficient, while the response weights of the DOFs in the third direction changes. Comparing the two curves shows that restricting the rotational vibration would result in more increase in the minimum clear distance required to avoid pounding, compared to the response in the perpendicular direction. Therefore, an optimized solution needs to be achieved using the evolutionary algorithms, if active control is an option for controlling coupled buildings with irregularity and SSI effects. Another major concern with coupling two buildings (Case-II) is that larger relative displacements may not be suitable for the application of MR dampers that are sensitive to shaft displacement and velocity. Due to such reasons, the application of MR dampers, as the connectors of two coupled buildings, is not recommended for this specific case study.

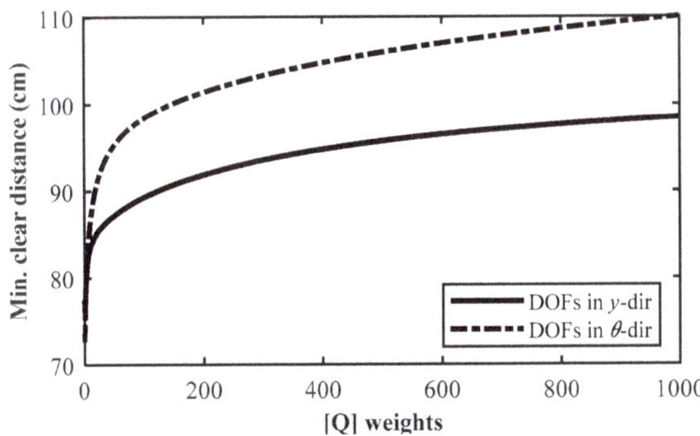

Figure 17. Minimum Clear distance required to avoid pounding using active control with respect to different response weights.

5.4. Advantages of SPC over Traditional FLC (Case-III)

The proposed SPC includes two FLCs in the main control flowchart, however, there are some advantages for the SPC, which a typical FLC may not be able to consider. Figure 18 shows the top floor displacement response of the five adjacent buildings under the bidirectional loads of the Northridge earthquake. The first row of the figure corresponds to the responses in the x-direction, which includes the buildings in the west, center and east (information flow in west-east direction). Similarly, the second row represents the responses in the y-direction, which shows the information flow among the buildings in the north, center and south. According to the information flow rule among the adjacent buildings for the proposed SPC, it is obvious that for regular adjacent buildings, the control algorithms are independents in two directions; however, for the current case (Case-III), controlling the vibrations in one direction influences the response in the other direction through torsional vibrations in the top floors. Since the stronger component of the Northridge earthquake is applied in the x-direction, the differences in the responses are obvious in this direction.

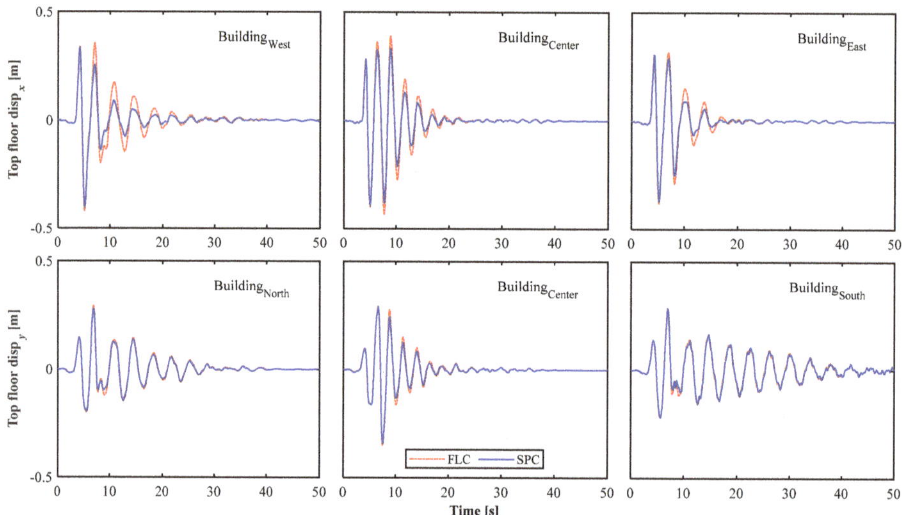

Figure 18. Comparison of the responses using fuzzy logic control (FLC) and SPC algorithms under the Northridge earthquake.

To investigate the performance of the proposed SPC algorithm in the case of damage, the buildings are excited under the Northridge earthquake. Based on the obtained lateral drift profiles, it is assumed that the first five floors of the building in the center lose the lateral stiffness in a way that the maximum and minimum damages in a range of 50%~25%, respectively, occur in the first and fifth floor. Besides, the control system of the building is shut down at the 5th second after the activation, when the drifts reach approximately maximum values. Therefore, the minimum damping capacities of the dampers are used for the rest of the excitation. Under this situation, it is expected that using the proposed SPC algorithm, the adjacent buildings will change their behavior to avoid pounding hazards. Figure 19 shows the response time-histories for the buildings with and without consideration of damage in the central building. It is clear that the top floor displacements of the building in the center are increased in both directions; however, the adjacent buildings behave differently, but with a reduction in the responses.

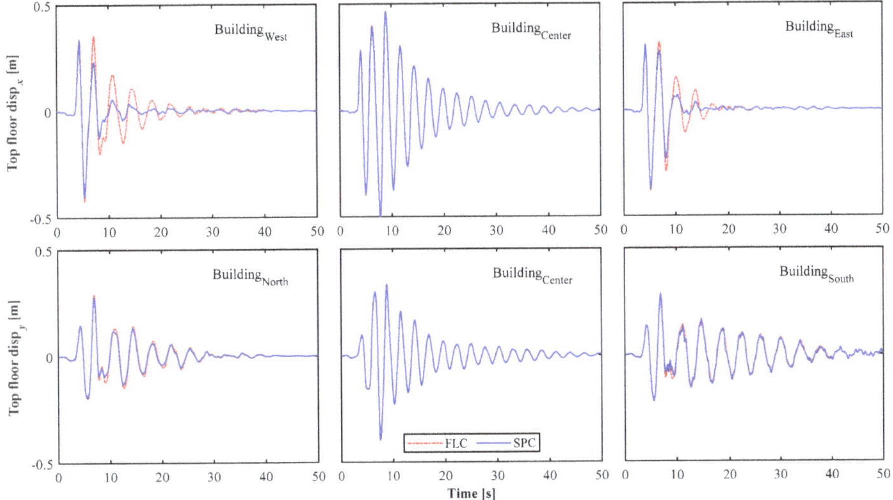

Figure 19. The response of the adjacent buildings considering damage in the central building using SPC.

6. Summary and Conclusions

The majority of the studies in seismic vibration control of three-dimensional tall buildings have not considered the irregularity and soil–structure interaction effects on the vibration response. In this paper, a novel bio-inspired seismic vibration control algorithm, called swarm-based parallel control (SPC), is proposed to use the advantages of fuzzy logic-based rules to consider the response of the individual adjacent building in determining the control forces for the group of buildings during extreme events, such as earthquake or wind loads. The main merits of the proposed SPC system can be summarized as:

1. Despite the other control approaches, such as using coupling devices, the need for the structural links to couple the two adjacent buildings, as well as the complexity, can be eliminated.
2. Each building can sense and consider the responses of the adjacent building in determining the optimal control force of semi-active devices; therefore, in the case of damage in a building, the other adjacent buildings update the control forces, accordingly.
3. The proposed SPC can be modified for individual buildings according to the deployed control strategy and devices.

To address the above-mentioned advantages of the proposed controller, numerical simulations are carried out under the seismic loads of seven historic earthquakes. Vibration responses of an irregular 15-story building are studied considering the soil–structure interaction (SSI) effect, as well as the effectiveness of different control methods (Case-I). A sensitivity analysis is also considered to study the performance of each control approach, considering uncertainties in estimating the soil and structural properties. In addition, the idea of coupling two adjacent regular-irregular buildings is discussed using an ideal active control, regardless of the control device type, and the limitations are highlighted (Case-II). It should be noted that for the active control system, the results are obtained based on the LQR algorithm to have a benchmark result for comparison, and the ideal conditions are assumed for either method. Finally, five adjacent buildings with different dynamic properties and no structural connections, considering the SSI effects, are modeled to verify the performance of the proposed SPC algorithm (Case-II).

The results prove that the proposed method has the potential to be paid attention in future developments as part of the bigger concept of the Internet-of-Things (IoT) and smart structures.

Author Contributions: Conceptualization, M.A.; methodology, M.A.; software, M.A. and A.M.Y.; validation, M.A.; formal analysis, M.A.; investigation, M.A. and A.M.Y.; resources, M.A.; data curation, M.A. and A.M.Y.; writing—original draft preparation, M.A.; writing—review and editing, M.A. and A.M.Y.; visualization, M.A.; supervision, M.A. All authors have read and agreed to the published version of the manuscript.

Funding: This project did not receive funding.

Acknowledgments: This study was initiated in 2018 based on Azimi [32], and the authors acknowledge and highly appreciate Zhibin Lin's academic support in the early stages of this study, from North Dakota State University.

Conflicts of Interest: The authors declare no conflict of interest.

References

1. Mosleh, A.; Rodrigues, H.; Varum, H.; Costa, A.; Arêde, A. Seismic behavior of RC building structures designed according to current codes. *Structures* **2016**, *7*, 1–13. [CrossRef]
2. Azimi, M. Stiffeners Effect on Seismic Performance of I-Beam to Double-I Built-up Column Connection. Master's Thesis, Iran University of Science and Technology, Tehran, Iran, November 2011.
3. Amiri, G.G.; Azimi, M.; Darvishan, E. Retrofitting I-beam to double-I built-up column connections using through plates and T-stiffeners. *Sci. Iran.* **2013**, *20*, 1695–1707.
4. Lin, J.L.; Tsai, K.C. Seismic analysis of two-way asymmetric building systems under bi-directional seismic ground motions. *Earthq. Eng. Struct. Dyn.* **2008**, *37*, 305–328. [CrossRef]
5. Mashayekhi, M.; Santini-Bell, E. Three-dimensional multiscale finite element models for in-service performance assessment of bridges. *Comput. Aided Civ. Inf.* **2019**, *34*, 385–401. [CrossRef]
6. Mashayekhizadeh, M. Fatigue Assessment of Complex Structural Components of Steel Bridges Integrating Finite Element Models and Field-Collected Data. Ph.D. Thesis, University of New Hampshire, Durham, NH, USA, 2018.
7. Mohammadi, R.K.; Nasri, A.; Ghaffary, A. TADAS dampers in very large deformations. *Int. J. Steel Struct.* **2017**, *17*, 515–524. [CrossRef]
8. Ghaffary, A.; Karami Mohammadi, R. Framework for virtual hybrid simulation of TADAS frames using opensees and abaqus. *J. Vib. Control.* **2018**, *24*, 2165–2179. [CrossRef]
9. Amini, F.; Samani, M.Z. A wavelet-based adaptive pole assignment method for structural control. *Comput. Aided Civ. Inf.* **2014**, *29*, 464–477. [CrossRef]
10. Uz, M.E.; Hadi, M.N.S. Optimal design of semi active control for adjacent buildings connected by MR damper based on integrated fuzzy logic and multi-objective genetic algorithm. *Eng. Struct.* **2014**, *69*, 135–148. [CrossRef]
11. Abdeddaim, M.; Ounis, A.; Djedoui, N.; Shrimali, M.K. Pounding hazard mitigation between adjacent planar buildings using coupling strategy. *J. Civ. Struct. Health Monit.* **2016**, *6*, 603–617. [CrossRef]
12. Chandiramani, N.K.; Motra, G.B. Lateral-torsional response control of MR damper connected buildings. In Proceedings of the ASME 2013 International Mechanical Engineering Congress and Exposition, San Diego, CA, USA, 15–21 November 2013.
13. Bozorgvar, M.; Zahrai, S.M. Semi-active Seismic Control of Buildings Using MR Damper and Adaptive Neural-fuzzy Intelligent Controller Optimized with Genetic Algorithm. *J. Vib. Control* **2019**, *25*, 273–285. [CrossRef]
14. Kaveh, A.; Dadras Eslamlou, A. An efficient two-stage method for optimal sensor placement using graph-theoretical partitioning and evolutionary algorithms. *Struct. Control Health* **2019**, *26*, e2325. [CrossRef]
15. Kaveh, A.; Dadras, A. Structural damage identification using an enhanced thermal exchange optimization algorithm. *Eng. Optimiz.* **2018**, *50*, 430–451. [CrossRef]
16. Cimellaro, G.P.; Lopez-Garcia, D. Algorithm for design of controlled motion of adjacent structures. *Struct. Control Health* **2011**, *18*, 140–148. [CrossRef]
17. Anajafi, H.; Poursadr, K.; Roohi, M.; Santini-Bell, E. Effectiveness of Seismic Isolation for Long-period Structures Subject to Far-field and Near-field Excitations. *Front. Built Environ.* **2020**, *6*, 24.

18. Roohi, M.; Hernandez, E.M.; Rosowsky, D. Nonlinear Seismic Response Reconstruction and Performance Assessment of Instrumented Wood-frame Buildings—Validation using NEESWood Capstone Full-Scale Tests. *Struct. Control Health* **2019**, *26*, e2373. [CrossRef]
19. Hernandez, E.; Roohi, M.; Rosowsky, D. Estimation of element-by-element demand-to-capacity ratios in instrumented SMRF buildings using measured seismic response. *Earthq. Eng. Struct. Dyn* **2018**, *47*, 2561–2578.
20. Cruz, E.F.; Cominetti, S. Three-dimensional buildings subjected to bi-directional earthquakes. Validity of analysis considering uni-directional earthquakes. In Proceedings of the 12th World Conference on Earthquake Engineering, Auckland, New Zeland, 30 January–4 February 2000.
21. Heo, J.S.; Lee, S.K.; Park, E.; Lee, S.H.; Min, K.W.; Kim, H.; Jo, J.; Cho, B.H. Performance test of a tuned liquid mass damper for reducing bidirectional responses of building structures. *Struct. Des. Tall Spec.* **2009**, *18*, 789–805. [CrossRef]
22. Yanik, A.; Aldemir, U.; Bakioglu, M. A new active control performance index for vibration control of three-dimensional structures. *Eng. Struct.* **2014**, *62*, 53–64. [CrossRef]
23. Nigdeli, S.M.; Boduroğlu, M.H. Active Tendon Control of Torsionally Irregular Structures under Near-Fault Ground Motion Excitation. *Comput. Aided Civ. Inf.* **2013**, *28*, 718–736. [CrossRef]
24. Wu, W.H.; Wang, J.F.; Lin, C.C. Systematic assessment of irregular building–soil interaction using efficient modal analysis. *Earthq. Eng. Struct. Dyn.* **2001**, *30*, 573–594. [CrossRef]
25. Wolf, J.P. Soil-structure-interaction analysis in time domain. *Nucl. Eng. Des.* **1989**, *111*, 381–393. [CrossRef]
26. Lin, C.-C.; Chang, C.-C.; Wang, J.-F. Active control of irregular buildings considering soil–structure interaction effects. *Soil. Dyn. Earthq. Eng.* **2010**, *30*, 98–109. [CrossRef]
27. Farshidianfar, A.; Soheili, S. Ant colony optimization of tuned mass dampers for earthquake oscillations of high-rise structures including soil–structure interaction. *Soil. Dyn. Earthq. Eng.* **2013**, *51*, 14–22. [CrossRef]
28. Alavi, A.H.; Hasni, H.; Lajnef, N.; Chatti, K.; Faridazar, F. An intelligent structural damage detection approach based on self-powered wireless sensor data. *Automat. Constr.* **2016**, *62*, 24–44. [CrossRef]
29. Ashtiani, R.S.; Little, D.N.; Rashidi, M. Neural network based model for estimation of the level of anisotropy of unbound aggregate systems. *Transp. Geotech.* **2018**, *15*, 4–12. [CrossRef]
30. Roohi, M.; Hernandez, E.M. Performance-based Post-earthquake Decision-making for Instrumented Buildings. *arXiv* **2020**, arXiv:2002.11702.
31. Gerist, S.; Maheri, M.R. Multi-stage approach for structural damage detection problem using basis pursuit and particle swarm optimization. *J. Sound. Vib.* **2016**, *384*, 210–226. [CrossRef]
32. Azimi, M. Design of Structural Vibration Control Using Smart Materials and Devices for Earthquake-Resistant and Resilient Buildings. Master's Thesis, North Dakota State University, Fargo, ND, USA, 2017.
33. Cheng, F.Y.; Jiang, H.; Lou, K. *Smart Structures: Innovative Systems for Seismic Response Control*; CRC Press: Boca Raton, FL, USA, 2008.
34. Nazarimofrad, E.; Zahrai, S.M. Seismic control of irregular multistory buildings using active tendons considering soil–structure interaction effect. *Soil. Dyn. Earthq. Eng.* **2016**, *89*, 100–115. [CrossRef]
35. Chopra, A.K. *Dynamics of Structures: Theory and Applications to Earthquake Engineering*; Prentice Hall: Upper Saddle River State, NJ, USA, 2011.
36. Azimi, M.; Rasoulnia, A.; Lin, Z.; Pan, H. Improved semi-active control algorithm for hydraulic damper-based braced buildings. *Struct. Control Health* **2017**, *24*, e1991. [CrossRef]
37. Azimi, M.; Pan, H.; Abdeddaim, M.; Lin, Z. Optimal Design of Active Tuned Mass Dampers for Mitigating Translational–Torsional Motion of Irregular Buildings. In Proceedings of the 7th International Conference on Experimental Vibration Analysis for Civil Engineering Structures, San Diego, CA, USA, 12–14 July 2017; pp. 586–596.
38. Azimi, M.; Pekcan, G. Structural Health Monitoring Using Extremely Compressed Data through Deep Learning. *Comput. Aided Civ. Inf.* **2019**. [CrossRef]
39. Kaveh, A.; Bakhshpoori, T.; Azimi, M. Seismic optimal design of 3D steel frames using cuckoo search algorithm. *Struct. Des. Tall Spec.* **2015**, *24*, 210–227. [CrossRef]
40. Dyke, S.J.; Spencer, B.F.; Sain, M.K.; Carlson, J.D. Phenomenological model of a magnetorheological damper. *J. Eng. Mech. ASCE* **1997**, *123*, 230–238.

41. Nugroho, P.W.; Li, W.; Du, H.; Alici, G.; Yang, J. An Adaptive Neuro Fuzzy Hybrid Control Strategy for a Semiactive Suspension with Magneto Rheological Damper. *Adv. Mech. Eng.* **2014**, *6*, 487312. [CrossRef]
42. Hu, Y.; Liu, L.; Rahimi, S. Seismic vibration control of 3D steel frames with irregular plans using eccentrically placed MR dampers. *Sustainability* **2017**, *9*, 1255. [CrossRef]

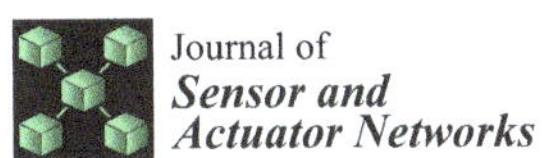
Journal of
Sensor and
Actuator Networks

Review

Emerging Trends in Optimal Structural Health Monitoring System Design: From Sensor Placement to System Evaluation

Robert James Barthorpe * and Keith Worden

Dynamics Research Group, Department of Mechanical Engineering, University of Sheffield, Mappin Street, Sheffield S1 3JD, UK; k.worden@sheffield.ac.uk

* Correspondence: r.j.barthorpe@sheffield.ac.uk

Received: 31 March 2020; Accepted: 24 June 2020; Published: 1 July 2020

Abstract: This paper presents a review of advances in the field of Sensor Placement Optimisation (SPO) strategies for Structural Health Monitoring (SHM). This task has received a great deal of attention in the research literature, from initial foundations in the control engineering literature to adoption in a modal or system identification context in the structural dynamics community. Recent years have seen an increasing focus on methods that are specific to damage identification, with the maximisation of correct classification outcomes being prioritised. The objectives of this article are to present the SPO for SHM problem, to provide an overview of the current state of the art in this area, and to identify promising emergent trends within the literature. The key conclusions drawn are that there remains a great deal of scope for research in a number of key areas, including the development of methods that promote robustness to modelling uncertainty, benign effects within measured data, and failures within the sensor network. There also remains a paucity of studies that demonstrate practical, experimental evaluation of developed SHM system designs. Finally, it is argued that the pursuit of novel or highly efficient optimisation methods may be considered to be of secondary importance in an SPO context, given that the optimisation effort is expended at the design stage.

Keywords: sensor placement optimisation; structural health monitoring; damage identification; mutual information; evolutionary optimisation

---

1. Introduction

The broad aim of Structural Health Monitoring (SHM) is to be able to objectively quantify the condition of an engineered structure on the basis of continually or periodically observed data, enabling incipient faults that may ultimately lead to failure of component or system to be detected at an early stage. These monitoring outcomes may be subsequently used to inform management of the structure, including decisions on how and when to deploy corrective actions and maintenance. Achieving high quality outcomes from the SHM process critically depends upon the quality of information gathered from the measurement system deployed upon the structure, which typically takes the form of a network of acceleration and/or strain transducers and appropriate acquisition hardware. Thus, the question of where to place the available sensors is key to achieving these desired outcomes. In most scenarios it is impractical to instrument every location of interest within a given structure. This is of particular concern in applications where the costs that are associated with deploying and maintaining a sensor network are high, for example in the aerospace sector where the impact of any mass addition on fuel efficiency must be considered. The desire to maximise the value of information returned by a reduced sensor set has led to the development of formal methods for sensor placement optimisation (SPO) for damage identification. As the size and complexity of the structures considered for monitoring has increased—with typical target applications ranging from aircraft to offshore wind turbines and large civil structures—so has the desire for effective methods for evaluating and optimising the design of SHM systems. The advent of cost effective and reliable wireless sensor networks adds a further consideration on this front.

This paper acts as an update on an earlier manuscript [1] and seeks to present the aims of SPO in an SHM context, set out the commonly adopted frameworks for tackling this problem and provide a critical review of the literature in this area. Most importantly, the particular contribution of this paper is to identify the key emerging topics that represent directions for future study. Given the wealth of literature in this area, it would be unwise to claim that this review is truly comprehensive; the quantity of (often closely related) works is simply too great. Instead, the paper is structured in order to highlight key themes within the literature and, consequently, demonstrate interrelations between approaches and avenues to novelty for researchers in this area.

A number of review articles exist on the topic of SPO in structural dynamics. Yi and Li [2] presents an overview of cost functions and optimisation methods focusing on civil infrastructure applications. A recent review article by Ostachowicz et al. [3] provides an update on the literature in this area with emphasis placed on methods appropriate to guided-wave SHM. A particular focus of the review in ref. [3] was to describe the numerous metaheuristic methods that have been applied in an SPO context; these are critiqued in Section 4.2. While a degree of overlap between reviews of this type is inevitable, a key distinguishing element of the current contribution is the emphasis on the identification of emerging methods that may allow the design of robustly-optimised SHM systems for which reliable a priori measures of expected benefit may be determined.

The layout of the paper is as follows. Section 2 provides an overview of the SPO problem as it applies to SHM. Section 3 introduces the cost functions. In Section 4, the optimisation methods that have been applied to the SPO task are summarised. Emerging trends are identified in Section 5 prior to concluding comments that are presented in Section 6.

2. Overview of the Sensor Placement Optimisation Problem

In a structural dynamics context, sensor placement optimisation first emerged for tasks such as the modal analysis of structures. Prior to the adoption of formal techniques, achieving successful modal analysis outcomes was highly reliant upon the experience and insight of the dynamicist conducting the test. By making use of practical steps, such as prioritising expected antinode location and avoiding expected nodes, it would be possible to create acceptable ad hoc sensor distributions that largely served well for distinguishing and identifying modes of non-complex structures. In cases where resource allowed, it would have been possible to trial different sensor configurations experimentally prior to adopting a final layout. However, as the complexity of the structures being considered has grown, it has become increasingly common practice to formalise the sensor placement problem as one of optimisation, with a cost function derived and subsequently minimised while using an appropriately chosen optimisation technique. The data that are used to conduct this optimisation may be drawn directly from experiment or, more commonly, may be based on the predictions of a numerical model. In parallel to this move to more formal optimisation methods has been the growth of research into SHM systems. While early SHM methods were largely based upon modal data (for example, changes in natural frequencies and modeshapes), recent years have seen a growth of methods making use of a far more diverse set of features, broadening the information sought from the SHM sensor network.

While the SPO problem may, in principal, be posed as one with a continuous outcome (e.g., returning the cartesian coordinates of individual sensor locations within a continuous space), the approach adopted almost universally in the reviewed literature is to pose the SPO task as one of combinatorial optimisation. This takes as it's starting point a candidate set of size n that contains all degrees of freedom (DOFs) that are available as sensor locations. The SPO task is to reduce this set to a smaller measurement set of size m that comprises only those DOFs to be employed as sensor locations. The number of combinations available for a given measurement sensor set size m is given by the binomial coefficient,

$$\binom{n}{m} = \frac{n!}{m!(n-m)!} \quad (1)$$

Searching for globally-optimal solutions becomes computationally demanding for anything other than cases with small n. Figure 1 provides an an illustrative example of an SPO application for SHM, and it serves to demonstrate this point. The objective of the SHM system in the presented example is to detect damage introduced into a composite aircraft wing structure via changes in the acceleration response spectra recorded at a set of measurement locations. For any given measurement configuration, a set of damage-sensitive features may be derived from the retained spectra. Representations of this feature set from both the damaged and undamaged state of the structure may subsequently be used to train a damage detector. In this example, a candidate set of 36 piezoelectric accelerometers are mounted on the upper surface of the wing, with the system excited using an electrodynamic shaker mounted towards the wingtip. Thus, the SPO task is to select the measurement sensor set that generates the most robustly discriminative damage detector.

Figure 1. A composite wing structure mounted in a free-free configuration. A candidate set of 36 accelerometers are mounted to its upper surface.

In the experimental example that is presented in Figure 1, the candidate set comprises $n = 36$ sensor locations. Setting the measurement set size to $m = 6$ sensors would result in 1.95×10^6 possible combinations to consider. Thus, arriving at an optimal measurement set in this scenario may become a computationally expensive task, and particularly so if the evaluation of a given measurement set involves tasks, such as training and testing a classifier. This issue grows in importance as n increases. To illustrate this, consider the common case that SPO for the structure shown in Figure 1 is conducted while using the predictions of finite element analysis (FEA) rather than data from the physical structure. If, for the purposes of illustration, $n = 1000$ nodal points were adopted as admissible candidate locations, the number of measurement set combinations for the same $m = 6$ scenario would grow to 1.37×10^{15}. This fact prompts interest in methods for efficient combinatorial optimisation, with an overview of applied techniques being given in Section 4.

3. Cost Functions

This section seeks to provide an overview of the key approaches that have been proposed for sensor placement in the structural dynamics context, building from foundational techniques, such as the effective independence (EI) and kinetic energy (KE) methods, towards more recent methods based upon classification outcomes that are more specifically tailored to the SHM task. The intention of this section is to highlight both the development arc of these methods and the key considerations that are to be made by the SHM practitioner when seeking to select, develop, and implement them.

3.1. Information Theory

Several of the cost functions introduced in this section are based upon information theory; this is natural, given that the essence of the SPO problem is to extract the most useful information possible given a restricted number of available channels (or, equivalently, measured DOFs in the SPO case). Prior to introducing the cost functions that have been applied for SHM systems, there is value in briefly reviewing relevant concepts.

3.1.1. Fisher Information

Fisher information offers a measure of the information that a sampled random variable X contains about an unknown parameter θ. Formally, Fisher information is based upon the score i.e. the gradient of the log-likelihood function, $\ln f(X;\theta)$, relating X, and θ. This gradient is usually expressed as the partial derivative of this function w.r.t. to θ. The Fisher information itself is then the variance of the score,

$$I(\theta) = \mathrm{E}\left[\left(\frac{\partial}{\partial\theta}\ln f(X;\theta)\right)\middle|\theta\right] \tag{2}$$

where E[.|.] denotes the conditional expectation, in this case with respect to θ. In the case that there are multiple unknown parameters $\boldsymbol{\theta} = [\theta_1 ... \theta_N]$, the Fisher information may be stated in matrix form with elements,

$$[I(\boldsymbol{\theta})]_{ij} = \mathrm{E}\left[\left(\frac{\partial}{\partial\theta_i}\ln f(X;\boldsymbol{\theta})\right)\left(\frac{\partial}{\partial\theta_j}\ln f(X;\boldsymbol{\theta})\right)\middle|\boldsymbol{\theta}\right] \tag{3}$$

Techniques that are based upon maximising the determinant of the Fisher information matrix (FIM) have been widely used as a basis for sensor placement optimisation. In these approaches, the available data, X, is typically taken to comprise the measurements available from a given sensor distribution. Various quantities may then be adopted for the unknown, sought parameters $\boldsymbol{\theta}$. For example, in the case of Effective Independence (described in Section 3.2.1) the quantities that are chosen for $\boldsymbol{\theta}$ are the elements of the modeshape matrix.

3.1.2. Mutual Information

Mutual information gives a measure of the mutual dependence between two random variables. The mutual information between two variables may be expressed as,

$$I(x,y) = \log_2\left[\frac{p_{X,Y}(x,y)}{p_X(x)p_Y(y)}\right] \tag{4}$$

where x and y are realisations of the jointly-distributed random variables X and Y, $p_{X,Y}$ is the joint probability density function of X and Y and p_X and p_Y are the marginal densities of X and Y.

For the SPO problem, mutual information may be interpreted as how much information one can "learn" about a sensor location from any another. If there are two sets of measurement locations, **A** and **B**, the amount of information that is learnt by sensor location a_i about location b_j is represented by the mutual information $I(a_i, b_j)$. In contrast to many applications of mutual information in machine learning, the approach that is taken in sensor placement optimisation is typically to minimise the mutual information between sensor locations, thus promoting independence. In the case that a_i and b_j are completely independent of one another, $I(a_i, b_j)$ drops to zero. A typical approach making use of sequential sensor placement (see Section 4.1) would involve **A** comprising a set of adopted measurement locations and **B** comprising a set of locations being considered for addition to it. The average mutual information between the adopted set of locations in **A** and each candidate for addition in **B** may be calculated by evaluating I for in a pairwise fashion and taking an average. The candidate location from set **B** that is determined to possess the lowest average mutual information with those in **A** is then adopted at each step.

3.1.3. Information Entropy

Information entropy offers a measure of the uncertainty in a model's parameter estimates. Optimal sensor placement might be achieved by minimising the change in the information entropy $H(D)$ given by,

$$H(X) = -\mathrm{E}\left[\ln \mathcal{L}(X|\theta)\right] \tag{5}$$

where θ is the uncertain parameter set, X represents the experimental test data, and $\mathrm{E}(.)$ denotes the expectation with respect to θ. A rigorous mathematical description is given in ref. [4], where it is shown that, for large datasets, the information entropy depends on the determinant of the FIM.

3.2. Modal Identification Based Cost Functions

The aim of the sensor placement task in a modal identification context is to select a measurement location set from a large, finite candidate set, such that the modal behaviour of the system remains as accurately represented as possible. Successful modal identification from a reduced sensor set requires three key decisions to be made, as discussed by Udwadia [5]:

1. Sensor quantity: how many sensors are required to enable successful modal identification?
2. Sensor placement: where should the available sensors be placed in order to best capture the required data?
3. Evaluation: how may the performance of the final, optimised sensor configuration be quantified?

Bounds on sensor quantity will generally be apparent from early in the test planning process. The lower bound is set by the requirement that for the modeshapes to be uniquely identifiable, the sensor quantity must be at least equal to the number of modes of interest. The upper bound will usually arise from resource limitations e.g., costs per channel or availability of equipment. Between these bounds, there may be some flexibility to consider the addition of sensors beyond the lower bound in order to allow for either greater ability to visualise modeshapes, or to provide a degree of robustness to sensor failure. Sensor placement and system evaluation are closely related and constitute the primary focus of this article. For the sensor placement step, the key elements are to adopt an appropriate performance measure/cost function (the focus of the present section); and then to select an appropriate method with which to optimise it (covered in Section 4). Evaluation of the performance of the optimal measurement set is an important, final gating activity prior to moving on to testing, enabling the expected performance of the proposed sensing system to be quantitatively evaluated and compared against test requirements. A distinction is drawn here between the evaluation methods used as part of the sensor placement optimisation step and those adopted for the final system evaluation step, as these may not necessarily be the same. It may, for example, be the case that SPO is conducted using a computationally inexpensive abstraction of the procedure applied at the final system evaluation step.

Note that the three decisions that are listed above will not necessarily be made in sequential steps; in general, there will be interplay and iteration between them. Also note that, while initially proposed in a modal identification context, these decisions are of no less importance when considering SPO for SHM.

3.2.1. Effective Independence (EI)

The effective independence (EI) method was introduced by Kammer [6] following earlier work by Shah and Udwadia [7]. The method makes use of the Fisher Information Matrix, with the modeshape matrix being adopted as the quantity of interest. Maintaining the determinant of the FIM leads to the selection of a set of sensor locations for which the modeshapes are as linearly independent as possible.

Central to the method is the EI distribution vector E_D, which calculates the contribution that is made by each individual sensor location to the rank of the prediction matrix E. E can only be full rank if the mode partitions resulting from a given measurement location set are linearly independent. The distribution vector E_D is defined as the diagonal of E,

$$E = [\Phi]\{[\Phi]^{\mathrm{T}}[\Phi]\}^{-1}[\Phi]^{\mathrm{T}} \tag{6}$$

where Φ represents the mass normalised modeshape of the system realised at all retained response locations. The algorithm is sequential in nature. At each step, the terms in E_D are ranked according to their contribution to the determinant of the FIM. The lowest ranked sensor location is identified and deleted from the candidate set, as are the corresponding elements in the modeshape matrix. The new, reduced sensor set is then re-ranked, and the process is repeated in a sequential manner until the desired number of sensors is obtained.

The work was extended in ref. [8] to consider the effect of model error on EI outcomes, and in ref. [9] to consider the effect of measurement noise. The application of a Genetic Algorithm (GA) based approach is presented in ref. [10]. An alternative approach to the EI technique is presented in ref. [11]. Instead of sequentially removing sensors from a candidate set, the sensor set is sequentially expanded until the required number of sensors is achieved. The method selects the sensor location that offers the greatest increase in the determinant of the FIM. This expansion can begin with a single sensor, and multiple locations can be added at each step of the iteration if required. This change in approach reduces the issue of computational expense associated with large candidate sets, and presents the possibility of specifying a desired (or existing) sensor configuration at the outset, to which further sensors may be optimally added. The approach was demonstrated for 27 target modes provided by a large FEA model of an aerospace vehicle. The results from the new expansion method and from a conventional EI reduction method were compared. From a candidate set of 29,772 locations, 389 were chosen as measurement locations, with agreement between the two approaches for all but 9 sensors. A significant saving in computational costs was made. A commonly observed drawback of the EI algorithm is that it does not penalise sensor locations that display low signal strength, thus making it susceptible to poor performance in noisy conditions.

Penny et al. [12] compared the performance of the EI algorithm against that of a scheme based upon classic Guyan reduction using an *a priori* FE model of a cantilever beam as a case study. The Guyan reduction approach proceeds by sequentially removing measurement locations at which the inertial forces are small in comparison to the elastic forces. The original motivation for Guyan reduction was to allow for large finite element models to be reduced to a smaller set of master DOFs prior to eigenvalue analysis, with the authors postulating that the behaviour that made particular DOFs suitable for adoption as master nodes would also make them good candidates for inclusion in an experimental measurement set. The measurement sets that were proposed by each method were compared using three evaluation measures: the size of the off-diagonal elements of the modal assurance criterion (MAC) matrix; the condition number of a singular value decomposition (SVD) of the modeshape matrix; and, the determinant of the FIM. While the off-diagonal elements of the AutoMAC matrix should drop to zero in the case that the modes are not correlated, the SVD measure offers a more explicit indication of the linear independence of the modeshape vectors. It was found that that the EI method performed better in the case that rigid body modes were present, while the Guyan reduction method worked well for cases where the structure was grounded and, thus, displayed no rigid body modes. However, the key conclusion was that, while both methods produced acceptable results, neither reached a global optimum due to the sequential deletion method adopted for optimisation. Sequential sensor placement schemes—-and alternatives—will be returned to in Section 4.

3.2.2. Modal Kinetic Energy (MKE)

The MKE method [13] seeks to prioritise sensors that are located at points of maximum kinetic energy for the modes of interest, on the assumption that these locations will maximise the observability of those modes. MKE indices are calculated for all of the candidate sensor locations i, as follows

$$MKE(i) = \sum_{r} \left[\phi_{ir} \sum_{j} \mathrm{M}_{ij} \phi_{jr} \right] \tag{7}$$

where ϕ is the target modeshape matrix, M is the mass matrix, r refers to the rth mode, and i, j to the ith and jth DOF, respectively. Thus, the modal displacements that are associated with each location are weighted by the corresponding component from the mass matrix M, providing a measure of the kinetic energy contribution of each location to the modes of interest.

The major advantage of the MKE method is that it promotes the placement of sensors in a location of high signal strength and so is less susceptible than EI methods to issues arising from low signal-to-noise ratios. However, this comes at the cost of not explicitly considering the independence of the returned modeshapes. Also note that the use of the mass matrix effectively limits the use of MKE to model-based approaches; down-selection from an experimental set is not an available option in this case.

Li et al. [14] explored the inherent mathematical connection between the EI and KE methods. It is shown that the first iteration of the EI method will always give the same result as the KE method, and that for the special case of a structure with an equivalent identity mass matrix the EI approach is an iterated version of the KE approach, with the reduced modeshapes re-orthonormalised at each iteration of the EI method. For cases with non-identity equivalent mass matrix the KE method incorporates the mass distribution that is associated with each candidate DOF as a weight, whereas the EI method is not mass dependent. Both methods are applied to data from the I-40 bridge in Albuquerque, New Mexico for three cases. From a candidate set of 26 locations, 25 locations are selected in Case 1, six are selected in Case 2, and eight are selected in case 3. For Case 1, the same sensor is selected for deletion by both methods, and the ranking sequence is identical. For Case 2, the same six sensors are selected and ranked in the same order by the two methods, but with different ranking values. In Case 3, there is agreement for seven of the eight required sensors, but discrepancies in the ranking values and ranking order.

3.2.3. Average Driving Point Residue (ADPR)

An alternative measure of the contribution of a given sensor location to the overall modal response of a system is given by the average driving point residue (ADPR) [15]. The intention of the ADPR is to provide a measure of the average response of the system at a given location to a broadband input that excites all modes of interest. The ADPR is calculated from data or FE predictions, as

$$ADPR(i) = \sum_{r} \frac{\phi_{ir}^2}{\omega_r} \tag{8}$$

where i is the candidate sensor location under consideration, $r = 1, \ldots, N$ are the modes of interest, ϕ_{ir} is the ith element of the rth modeshape, and ω_r is the rth modal frequency. The ADPR in the form stated above provides a measure of average modal velocity, although minor variations allow modal displacements or accelerations to be considered instead [16]. In common with the closely-related MKE method, an advantage of using the ADPR approach over EI methods is that it tends to select sensors in areas of high signal strength, although, once again, at the cost of sacrificing explicit consideration of modeshape independence. The key differentiator of the ADPR and MKE approaches is that the ADPR does not require knowledge of the mass matrix, which enables it to be more easily applied in purely experimental scenarios where a physics-based model is not available.

A minor adaptation of the EI approach was introduced by Imamovic [16]. In the Effective Independence Driving Point Residue (EI-DPR) approach, the EI and ADPR values for each candidate location i are combined to produce an EIDPR value

$$E_D^{EIDPR}(i) = E_D^{EI}(i)ADPR(i) \tag{9}$$

Combining the metrics in this way was found to promote the placement of sensors in regions of higher signal to noise ratio than for the EI method alone, while also resulting in comparatively uniform sensor configurations. The Eigenvector Product (EVP) is a further closely-related method [17]. As the name suggests, this metric involves simply taking the product of the eigenvector elements for a given candidate location across all modes of interest,

$$EVP(i) = \prod_r |\phi_{ir}| \tag{10}$$

where i is the candidate sensor location and ϕ_{ir} is the ith element of the rth modeshape. A maximum for this product is deemed to be a candidate measurement location.

3.2.4. Modeshape Sensitivity

Shi et al. [18] present one of the earliest sensor placement approaches specific to structural health monitoring. A structural damage localisation approach that is based on eigenvector sensitivity is adopted, and a sensor placement optimisation approach is developed that uses the same method. The FIM approach is applied to the sensitivity matrix that is to be used for damage localisation, with those degrees of freedom that provide the greatest amount of information for localisation retained. The prediction matrix in this case is given by

$$E = [\mathrm{F(K)}]\{[\mathrm{F(K)}]^{\mathrm{T}}[\mathrm{F(K)}]\}^{-1}[\mathrm{F(K)}]^{\mathrm{T}} \tag{11}$$

where F(K) represents a matrix of sensitivity coefficients of modeshape changes with respect to a damage vector comprising stiffness matrix changes. The sensor placement and damage localisation algorithms are demonstrated for numerical and experimental data, and they are shown to be effective in indicating probable locations of both single and multiple damage locations. For the experimental survey of an eight-bay, three-dimensional truss structure, a set of 20 optimised measurement locations is selected from a candidate set of 138 DOFs. Correlation between the analytical and test modeshapes gives MAC values of over 0.99 for the first five modes. These five modes are used for the successful localisation of damage represented by the unscrewing of truss elements.

Guo et al. [19] highlight the difficulties that may occur in solving the matrix proposed by Shi et al. [18]. The problem is reformulated as an objective function that is based on damage detection, to be solved while using a genetic algorithm. An improved genetic algorithm (IGA) is introduced to deal with some of the limitations of the simple GA. The results from the new algorithm are compared with those from the penalty function method and the forced mutation method for a numerical simulation. The numerical example studied is based upon the same six-bay, two-dimensional truss used in Shi [18]. The improved genetic algorithm is shown to converge in far fewer generations than the penalty function and forced mutation methods. Improved convergence characteristics and low sensitivity to the initial population used are also demonstrated.

3.2.5. Strain Energy Distribution

Hemez and Farhat [20] present a sensor placement study that is specific to damage detection based on strain energy distributions. The EI algorithm that is presented in ref. [6] is modified to allow the placement of sensors according to strain energy distributions. First, a strain energy matrix E is constructed,

$$E = [\Psi^{\mathrm{T}}\Psi] \tag{12}$$

where $\Psi = C^T$ and C represents a Cholesky decomposition of the stiffness matrix K i.e., $K = C^T C$. The elements of the distribution vector E_D that lies on the diagonal of E thus represents the contribution of each sensor location to the total strain energy recorded by the system. The proposed method proceeds in the same manner as the original EI algorithm, with the sequential deletion of locations from the candidate set until the desired number of sensors is reached. The performance of the EVP and EI methods is compared in ref. [20] for a damage detection case study on an eight-bay truss structure. The objective for the sensor network is to facilitate optimal model updating in the presence of damage. It was found that the damage detection performance of the updated model was sensitive to the chosen sensor distribution. While both of the algorithms returned measurement sets that are capable of detecting damage, the update that is based on the EI measurement set contained some inaccuracies when reporting the damage locations.

3.2.6. Mutual Information

In ref. [21], the selection of optimal sensor locations for impact detection in composite plates is approached while using a GA to optimise a fitness function based upon mutual information. The mutual information concept is used to eliminate redundancies in information between selected sensors, and rank them on their remaining information content. The method is experimentally demonstrated for a composite plate stiffened with four aluminium channels, with an optimal set of six sensors selected from a candidate set of 17. The selected locations are found to lie close to the stiffening components. It is concluded that this result is consistent with the stiffened regions being the most challenging for the assessment of impact amplitude; the deflections under impact in these regions are much smaller than those observed in the unconstrained centre of the plate.

Trendafilova et al. [22] use the concept of average mutual information to find the optimal distance between measurement points for the sensor network. An equally-spaced configuration is assumed, and the sensor spacing varied in order to minimise the average mutual information between measurement locations. The approach is tested for a numerical simulation of a rectangular plate using a previously-established damage detection and location strategy, while using damage probabilities as the assessment criteria. The approach is tested for three spacing values: an initial set containing 36 sensors, the optimal set given by the mutual information method and containing eight sensors, and a third set containing six sensors where the spacing between locations is greater than the calculated optimal value. The results show that the initial and optimally-spaced sets are equally successful in locating damage. The set with a greater spacing between sensors performs significantly less well.

3.2.7. Information Entropy

Papadimitriou et al. [23] present a statistical method for optimally placing sensors for the purposes of updating structural models that can subsequently be used for damage detection and localisation. The optimisation is performed by minimising information entropy, a unique measure of the uncertainty in the model parameters. The uncertainties are calculated while using Bayesian techniques and the minimisation realised using a GA. The method is tested using simplified numerical models of a building and a truss structure. The results show that the optimal sensor configuration depends on the parameterisation scheme adopted, the number of modes employed, and the type and location of excitation.

3.3. *SHM Classification Outcomes*

The dominant paradigm in SHM in recent years is that based upon statistical pattern recognition. This involves training statistical classification and/or regression algorithms in either a supervised or unsupervised manner. Thus, it is natural to consider classification outcomes as the basis for constructing cost functions. Formally, this requires the selection of metrics on classifier performance to be developed and evaluated using either the same data used for training (referred to as recall) or performance on an independent testing set [24]. Studies that adopt cost functions that are based on classification outcomes are reviewed in Section 4, alongside the optimisation methods applied.

For the simplest case of binary classification, metrics may be drawn from a confusion matrix of the form that is shown in Table 1.

Table 1. Exemplar confusion matrix for binary classification.

		Predicted Class	
		Positive	**Negative**
Actual Class	Positive	TP	FP
	Negative	FN	TN

In the confusion matrix, the quantities TP, FP, TN, and FN refer, respectively, to the number of True Positive, False Positive, True Negative, and False Negative classification outcomes that are observed for observations from a presented dataset. The principle quantities of interest that may be extracted from them are the probability of detection P_D (or true positive rate (TPR)), which is given by,

$$P_D = \text{TPR} = \frac{\text{TP}}{\text{TP} + \text{FN}} \tag{13}$$

and the probability of false alarm rate P_{FA} (or *false positive rate* (FPR)), which is given by,

$$P_{FA} = \text{FPR} = \frac{\text{FP}}{\text{FP} + \text{TN}} \tag{14}$$

A perfect classifier would return a P_D of 1 and a P_{FA} of 0.

In general terms, classifiers operate by computing a score that is related to class membership and then comparing this score to some predetermined classification threshold. Thus, the particular choice of threshold value is critical in controlling the trade-off between the true positive and false positive rates. The Receiver Operating Characteristic (ROC) is a useful tool for providing a comparison between the TPR and FPR for any given threshold level. By evaluating the ROC value, a discrete set of possible threshold values, data may be gathered with which to plot an ROC curve. This curve may be employed to (a) evaluate the discriminative ability of the data and (b) select a level for the threshold that is appropriate to the problem, and it is for the first of these functions that it is typically applied in an SHM SPO context.

An example of an ROC curve for a binary classifier is given in Figure 2. In this example, a "Negative" class label indicates that the structure is undamaged and a "Positive" class label indicates that damage has occurred. A classifier offering no discrimination would correspond to the line of no discrimination indicated on the diagonal. The further that the ROC curve for a given classifier lies above this line, the greater its discriminatory performance.

The area under the ROC curve (AUC) offers a further, quantitative method for summarising the level of discrimination offered by a classifier. The AUC returns a value in the range [0,1], where a value of 1 would indicate perfect classification performance. A classifier that assigned labels at random would return a value of 0.5, with its performance being as indicated by the line of no discrimination in Figure 2. The AUC might be interpreted as the probability that the considered classifier will rank a randomly selected positive instance higher than a randomly chosen negative instance [25].

Classification performance-based metrics may be extended to the localisation task, with an exemplar confusion matrix for a three class case presented in Table 2.

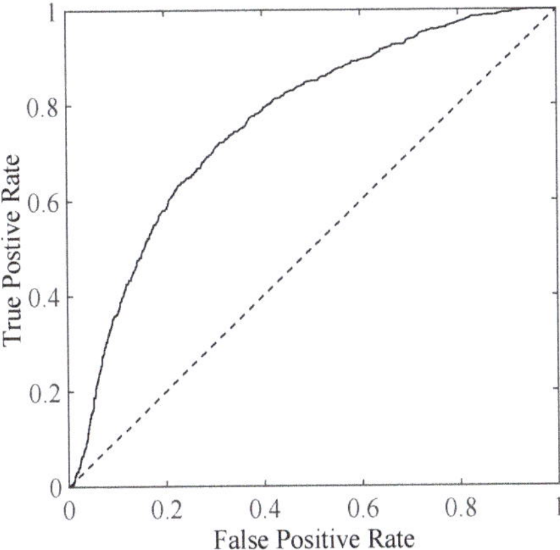

Figure 2. An example of a Receiver Operating Characteristic (ROC) curve for a moderately well-separated dataset. For comparison, the line of no discrimination is indicated by the dashed line along the diagonal.

Table 2. Exemplar confusion matrix for damage localisation.

		Predicted Location		
		A	**B**	**C**
Actual Location	A	TP_A	E_{BA}	E_{CA}
	B	E_{AB}	TP_B	E_{CB}
	C	E_{AC}	E_{BC}	TP_C

For damage localisation, the class labels (A, B, and C in the example) will typically refer to a discrete set of potential damage locations, with TP_A representing the number of true positive outcomes for class A, E_{BA} representing mislabelling of an observation from class A as being from class B and so on. The metric of most interest is the probability of correct classification (P_{corr}) over all K classes i.e., the sum of the diagonal quantities TP_k (with $k = 1...K$) divided by the total number of observations in the test set, which may be expressed as

$$P_{corr} = \frac{\sum_k TP_k}{\sum_k TP_k + \sum_{i \neq j} E_{ij}} \tag{15}$$

An alternative approach for defining metrics on classifier performance is to explicitly evaluate the performance of the proposed subset relative to that of the full set. Such a scheme is proposed in ref. [26] for a Neural Network classifier, where the normalised mean squared error (NMSE) between a desired (i.e., best possible) classifier response and that achieved by a classifier trained on a subset of measurement locations is used as the basis for a range of cost (alternatively "fitness") functions. The NMSE is given by,

$$\text{NMSE}(\hat{y}_i) = \frac{100}{N_T \sigma_i^2} \sum_{j=1}^{N_T} (y_i(j) - \hat{y}_i(j))^2 \tag{16}$$

where i represents the ith output neuron, N_T is the number of training sets indexed by j, and σ_i^2 is the variance of the output y_i. In ref. [26], it was proposed that the performance of the SHM system (a classifier trained using a subset of measurement points) may be judged on either the average value of the NMSE over the whole set of classifier outputs; or alternatively using the maximum of the set of output NMSEs.

Finally, cost functions that are specific to particular classifiers may be adopted. These include the margin of separation for Support Vector Machine (SVM) classification, for example.

3.4. Summary

The methods highlighted in this section follow a progression from those developed for the general sensor placement case (e.g., for the purposes of modal analysis), through to those that are specific to SHM. Two dominant groups of approaches are apparent: methods that are based upon variations on the Fisher information matrix (and other information metrics); and, methods based upon maximising classification outcomes. There does not yet appear to be strong consensus within the literature as to which specific approaches are the most appropriate for given scenarios, or an agreed set of selection criteria. However, it is apparent that the decision on which method to adopt in a given scenario is likely to be guided, to a large extent, by the information and resource available to the practitioner. A key element is the availability of damaged state data—a topic of general concern in SHM. Methods that are based on classification outcomes will, in general, require the availability of either experimentally-obtained representations of damaged state data or, more realistically, numerical model predictions of damaged state system response. Therefore, the quality and robustness of sensor placement outcomes will therefore be a function of the information one is able or prepared to gather and include within the analysis. These concepts are explored further in Section 5.

4. Optimisation Methods

The sensor placement optimisation task (whether for SHM purposes or otherwise) is typically approached as a combinatorial optimisation problem i.e., the aim is to select some subset of size m from a discrete, finite set of size n that optimises a chosen cost function. The approaches that are available for performing combinatorial optimisation of the cost functions introduced in Section 3 are largely drawn from two categories: sequential and metaheuristic methods. Where computationally feasible, it is deemed to be good practice to compare optimisation outcomes to those of an exhaustive search.

4.1. Sequential Sensor Placement Methods

Sequential sensor placement (SSP) methods offer a set of simple to implement and computationally efficient heuristic algorithms that are appropriate to the sensor placement task. There are two principal approaches: backward sequential sensor placement (BSSP) and forward sequential sensor placement (FSSP). Both of the methods proceed along similar lines, with the BSSP algorithm starting with a large sensor set and sequentially eliminating those sensor locations that contribute least to the objective function (as in the original EI study by Kammer [6]), and FSSP beginning with a small set and adding sensor locations that offer the greatest benefit (as explored for EI in ref. [11]). These approaches are discussed in detail in refs. [4,27], with the performance of the sub-optimal sensor distributions returned by the FSSP and BSSP method shown to provide a good approximation to the performance of the optimal sensor distribution on a numerical test case.

Stephan [28] proposes a method that acts as a hybrid of the FSSP and BSSP methods. The approach is demonstrated for modal test planning on the basis of a finite element model of an aircraft. Two steps are carried out within each iteration of the algorithm. First, a new sensor location is proposed, based upon its contribution to the Fisher information matrix. Next, the levels of redundancy between the newly proposed sensor and each sensor already within the proposed distribution is evaluated, with any that fall below a pre-set threshold value eliminated. The method is shown to perform well in comparison to the standard EI method, with the clustering of sensor locations reduced.

4.2. Metaheuristic Methods

In recent years, metaheuristic (and particularly evolutionary) optimisation methods have become manifest. Metaheuristic methods are stochastic in nature and they seek to find solutions that approach the global optima. There are a wide variety of such methods, with many inspired by processes occurring in nature, such as evolution or animal behaviour. These methods are given extensive coverage in ref. [3]. The general approach adopted by metaheuristic optimisation techniques is as follows:

1. generate an initial population (the First Generation) containing randomly-generated individuals. In an SPO context, each 'individual' is typically taken to comprise a sensor location subset;
2. evaluate the fitness of each individual within the population against a pre-defined cost function;
3. produce a new population by (a) selecting the best-fit individuals for reproduction and (b) breeding new individuals through crossover and mutation operations; and,
4. repeat steps (2) and (3) above until some stopping criteria is reached.

4.2.1. Genetic Algorithms (GA)

GAs offer a powerful class of metaheuristic method for the solution of search and optimisation problems. They are directly inspired by the concept of natural selection with candidate solutions being encoded as gene vector, usually in the form of a binary string. GAs have been extensively used in SPO applications. A typical formulation is for the gene vector to be of length n, with each bit within the vector indicating whether or not a sensor is placed at that location.

Yao et al. [10] presented one of the earliest applications of a GA for sensor placement. Developing from the standard EI formulation proposed in ref. [6], which makes use of a backward sequential approach, the determinant of the FIM is adopted as the fitness function for the GA. The approach is applied to SPO for modal identification with two case studies considered: a space structure and a photo-voltaic (PV) array. It was found that the GA marginally outperformed the EI algorithm for an $m = 10$ and $m = 20$ measurement set case for the space structure. However, the GA was found to be significantly more costly, incurring 30 times the computational cost for the 20-sensor case. Similar results were observed for the PV array. It was noted that the GA used may converge to a local minimum, highlighting a general issue with search methods of this type. A typical approach to overcome this is to repeat the optimisation multiple times with randomisation of the starting population.

Stabb & Blelloch [29] extended the application of GAs for SPO in a modal analysis context to two further fitness functions. The first of these is simple the MAC. The second is the cross-orthogonality matrix between the modeshapes of interest, partitioned according to the candidate sensor set and the associated Guyan-reduced mass matrix. The MAC approach was found to quicker thanks to the lower computational cost of each evaluation, but was also found to be the less accurate of the two, as the FE modes are only truly orthogonal with respect to the mass matrix. For each measure, an error function is computed by taking the absolute value of the difference with an "ideal" matrix: the identity in the case of the cross-orthogonality matrix, and the MAC from the full modeshapes in the case of the MAC. The GA is compared with the EI and MKE methods, as well as the iterative Guyan-reduction method that is introduced in Penny et al. [12]. For a case study that is based upon an FE model of a, it was found that the best solution of all the alternatives tested was achieved using the method incorporating the Guyan-reduced mass matrix and GA optimisation, with the GA seeded using the results of a traditional EI approach. This method produces accurate modal frequencies for four out of six modes, whereas the traditional EI and KE methods performed poorly.

In one of the earliest studies on sensor placement specific to an SHM problem, Worden et al. [30] use a neural network (NN) for the location and classification of faults and a GA to determine an optimal (or near optimal) sensor configuration. The NN is trained using modeshape curvatures provided by an FE model of a cantilever plate. Faults in the plate were simulated by "removing" groups of elements in the model. The probabilities of misclassification for the different damage conditions are obtained from the neural network, and the inverse of this measure is employed as a fitness function for the GA. A set of 20 locations were selected as a candidate sensor location set, and the GA optimisation compared to the results of two heuristic approaches: a deletion strategy and an insert strategy. The results of this preliminary study showed the GA to outperform both heuristic methods.

In Worden and Staszewski [31], a NN/GA approach was used to place sensors for the location and quantification of impacts on a composite plate. In this experimental survey, the strain data are recorded for impacts applied to a wingbox structure. An artificial neural network is trained to locate the point of impact, and a second NN employed to estimate the impact force. A GA was used to select an optimal set of three measurement locations from the candidate set of 17 used in the test, while using the estimation of the impact force provided by the neural network as the parameter to be maximised. As this parameter is dependent upon the starting conditions used to train the neural network, the training was run six times to allow for comparison between GA results and those of an exhaustive search. For each run, the GA found the same optimal set as the exhaustive search.

Coverley and Staszewski [32] presented an alternative damage location method combining classical triangulation procedures with experimental wave velocity analysis and GA optimisation. The triangulation approach assumes that three different angles for wave propagation directions from an unknown impact location are available. The distance between the sensor and the unknown impact position is calculated using the arrival times and the velocities of the propagating elastic waves. While the method was shown to perform well, a limitation of the triangulation approach is the need to estimate the velocities of the propagating elastic waves—something that is challenging in anisotropic materials. De Stefano et al. [33] adopted a trilateration approach that attempts to overcome these issues. Trilateration does not require the measurement of angles of approach, instead being based purely on the measurement of distances. The resulting system of equations is non-linear in the solution parameters, thus requiring a non-linear least-squares algorithm in order to calculate a solution. The major advantage of the trilateration method is that a single NN might be trained prior to optimisation taking place, thus offering a major reduction in computational cost in comparison to schemes where a NN must be trained at each iteration within the optimisation algorithm.

One of the key issues discussed in ref. [26] relates to appropriate methods for encoding sensor locations. The simplest way to encode an m-sensor distribution is in an n-bit binary vector, where n is the size of the candidate location set and each bit switches on if the corresponding sensor is present. However, this is suboptimal as the distribution for the sensor sets will follow a binomial distribution, and will thus be very sharply peaked at $n/2$ sensors. If n is large, randomly-generated distributions will almost certainly have close to $n/2$ sensors, regardless of the value of m. This means that crossover or mutation will usually be destructive unless protected operations are defined. The simplest alternative is to use an integer-valued GA where there are m integers, each indexing a sensor in the candidate set. Protected operators may still be convenient, or a penalty function can be used to eliminate distributions with repeats.

4.2.2. Simulated Annealing (SA)

In ref. [26], simulated annealing was one of three sensor placement methods compared using an FE model of a cantilever plate. Fault diagnosis was performed while using a neural network trained on modeshape and curvature data generated using the numerical model, resulting in the structure being classified as either faulted or not faulted. The objective function that was chosen for the SA algorithm was the probability of misclassification. The SA method was shown to perform impressively, showing a slight improvement on a comparable GA approach when 10 sensors were selected from a candidate set of 20. The SA algorithm yielded a four-sensor distribution with a 99.5% probability of correct classification, and a three-sensor distribution that was only slightly different to that found by exhaustive search.

4.2.3. Ant Colony and Bee Swarm Metaphors

The same neural network problem as in ref. [31] is investigated using a different optimisation technique in Overton and Worden [34]. Here, an ant colony metaphor was used to place sensors based upon the fitness functions that are generated by the neural network, and the results compared with those from the GA and exhaustive search. For the selection of a three-sensor distribution from a set of 17, the ant colony algorithm is shown to be faster than the GA while producing a solution that is amenable to that from the exhaustive search. The algorithm also appeared to outperform the GA when a six-sensor distribution was sought, although entirely fair comparison between the approaches was not possible. As the size of the sensor network increases, it appears that the selected distributions become less intuitive.

In common with other NN-based approaches, it is noted that the objective function adopted in ref. [34] is a random variable conditioned on the initial weights given to the network. Scott and Worden [35] describe the development of a bee swarm algorithm, again via application to the NN-based objective function proposed in ref. [31]. The bee swarm algorithm was shown to perform well for an initial, constrained set of neural network parameters. However, the key development was to extend the algorithm in order to include optimisation of the neural network parameters alongside the sensor distributions. In this set up the bee swarm algorithm was shown to significantly improve accuracy, outperforming both the ant colony algorithm and the genetic algorithm in cases where performance could be directly compared. It was noted that the nature of the NN-based objective function was well-suited to the bee swarm algorithm, which only requires there to be some similarities—however vague—between "good" solutions in order to substantially outperform a random search.

4.2.4. Mixed Variable Programming (MVP)

In Beal et al. [36], optimal sensor placement is formulated as a mixed variable programming (MVP) problem. As observed in ref. [26], a typical formulation for the SPO process is to treat each sensor location as a binary variable (the sensor is either placed at the given location, or not), with the resulting optimisation being a binary mixed integer non-linear programming problem. A difficulty that arises from this mixed integer formulation is that it scales poorly, becoming computationally demanding as the candidate sensor set becomes large. The proposed alternative is a mixed variable programming scheme, where each sensor is assigned a categorical variable from a predefined list indicating where it is to be located. The MVP framework is sufficiently general that it may be applied a broad set of objective functions. The objective function used for demonstration in ref. [36] is the relative error between the stiffness changes indicated by the full and reduced measurement sets. The approach is demonstrated using a numerical model of a 20-DOF system. Given that damage had occurred at specified masses, optimal configurations containing three-, four-, and five-sensors were found for three damage cases. A convergence analysis is presented that indicates strong results for optimisation problems with one continuous variable, indicating that the approach is sufficiently general for a variety of SPO problems.

4.3. Summary

Given the extensive literature in this area, it has become increasingly challenging for authors in this area to demonstrate true novelty of approach. This issue is exacerbated in cases where there is no principled comparison either to other methods or to best possible outcomes. The best possible outcome might be interpreted either as the outcome achievable if all data are available (i.e., the full sensor set may be used); or, the best possible outcome achievable for a given sensor subset size, evaluated via exhaustive search. In either interpretation, the best possible performance represents an upper bound on system performance.

Also noted is that the connection between the problem to be solved (minimisation of a particular cost function relating to sensor placement) and the optimisation method chosen is given little explicit consideration in many of the papers that appear in the SPO literature. The "no free lunch" theorems popularised by Wolpert and Macready [37] establish that for any given optimisation algorithm, over-performance for one class of problems will be offset by under-performance for another class. Thus, for uniformly distributed problem classes (i.e., the likelihood of being faced with a "simple" optimisation task is the same as being presented with a "hard" task), the performance of any given optimisation algorithm will be equal to that of any other when averaged across all classes. It is only by making use of *a priori* understanding of the particular problem at hand that overall improved performance can be achieved.

5. Emerging Trends and Future Directions

This section seeks to identify research directions that have the potential to make a substantial contribution to the problem of system design for SHM. It is apparent from the quantity of papers available in the literature, of which only a subset are reviewed in Sections 3 and 4, where sensor placement optimisation is, in some senses, a mature field. Optimisation methods applicable to the combinatorial subset selection task have, in particular, been extensively explored. While there may still be some scope for tailoring optimisation methods to particular approaches in the pursuit of computational efficiency, there appears general agreement that for any given appropriately-posed problem, an acceptable solution will be provided by the methods that are covered in Section 4. There is arguably more scope for novelty in regard to cost functions, particularly with regard to tailoring cost functions to specific SHM systems. However, the avenue for real progress appears to be in directly addressing the trade offs that are required to make a decision on whether or not to implement an SHM system. There is agreement here with the perspectives expressed in ... and more recently of ref. [38] that the focus of SHM research should—at this stage in its development—be on moving towards industrial deployment of monitoring systems.

Key among the trends identified and discussed are applications of methods that are drawn from decision theory (Section 5.1) and the quantification of overall SHM system value (Section 5.2). A comparatively recent trend is to seek to make the sensor network robust to confounding factors, including: (1) benign effects driven by Environmental and Operational Variables (EOVs); (2) uncertainties and errors in any contributing mathematical or numerical models; and, (3) sensor and sensor network failure. The emerging literature in each area is briefly reviewed in Sections 5.3 and 5.4. Finally, methods that consider robustness to sensor or sensor network failure are considered (Section 5.5), as are the opportunities for gathering full-field, experimental candidate set data via laser vibrometry (Section 5.6). Several of the reviewed papers do not explicitly cover sensor placement optimisation for SHM, but introduce ideas or methods that are relevant to further development in this field.

5.1. From Classifiers to Decisions

A key emerging trend within the research literature is to move from the optimisation of classification outcomes to the optimisation of decision outcomes. Flynn and Todd [39] present perhaps the most influential recent study in this field, setting out a Bayesian risk minimisation framework for optimal sensor placement. This framework enables the costs associated with different actions to be included in the decision-making process, enabling a design that minimises risk under uncertainty to be found. Here, design, is taken to incorporate sensor placement, feature selection, and the setting of localised detection thresholds. This essentially casts damage identification as a binary (or more generally M-ary) hypothesis testing problem, weighted by the cost of making a correct/incorrect decision. The framework proposed is general and, in principle, could be applied to multisite damage cases with costs that are associated with all outcomes. However, through a series of assumptions the approach is simplified to the consideration of binary classifiers that are associated with local health states, with the structure discretised into K spatial regions, indexed by $k = 1...K$. Figure 3 shows a schematic illustration of a discretised structure. For each region, a local detection threshold $T[k] > \gamma[k]P(h_{k0})/P(h_{k1})$ is established, where $\gamma[k]$ represents the classification cost ratio, h_{kj} represents the true local damage state (with $j = 0$ and $j = 1$ denotes undamaged and damaged states, respectively), and $P(h_{kj})$ represents prior probabilities on the local damage states.

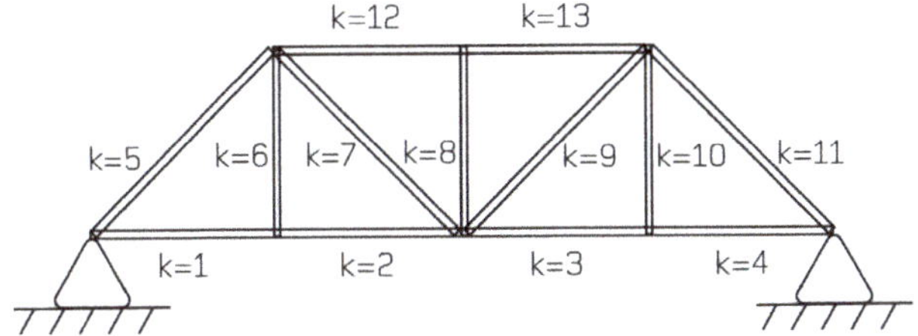

Figure 3. Schematic example of a discretised structure.

Given these assumptions, the global probability of detection P_D is given by

$$P_D = \sum_{k=1}^{K} \frac{P(d_{k1}|h_{k1})P(h_{k1})}{\sum_{k=1}^{K} P(h_{k1})} \tag{17}$$

where d_{kj} represents the event of deciding that h_{kj} is the local damage state in region k and the local detection rate is given by,

$$P(d_{k1}|h_{k0}) = P_r(T[k] > \gamma[k]|h_{k1}) \tag{18}$$

Similarly, the global probability of false alarm rate P_{FA} is given by,

$$P_{FA} - \sum_{k=1}^{K} \frac{P(d_{k1}|h_{k0})P(h_{k0})}{\sum_{k=1}^{K} P(h_{k1})} \tag{19}$$

with the local false alarm rate given by,

$$P(d_{k1}|h_{k1}) = P_r(T[k] > \gamma[k]|h_{k0}) \tag{20}$$

While the particular case study in ref. [39] focuses on active sensing, the framework presented is appropriate for general application. The advantage of the assumptions made is that closed-form analytical expressions may be derived for P_D and P_{FA} as a function of the costs, prior probabilities of damage and local threshold damage threshold values. The optimisation task is then to maximise P_D for a given allowable false alarm rate P_{FA}; or alternatively, minimise P_{FA} for a given target P_D. The key outstanding question that remains is how best to establish the feature distributions that are associated with each health state of interest.

5.2. Quantifying the Value of SHM Systems

Going further, Thõns [40] presents a study considering the value of an SHM system as the quantity of primary concern. A framework for quantification of the value of SHM in the context of the structural risk and integrity management is presented. This approach seeks to directly address the core decision to be made by an asset manager: can the installation of the proposed system be justified? The approach adopted builds, as in ref. [39], on the notion of expected utilities associated with available decisions. The framework is demonstrated for a structural system undergoing a process of fatigue damage, with decision processes modelled in detail along with associated uncertainties. It is demonstrated that the value of information provided by an SHM system varies significantly between contexts, with the relative value being particularly strongly influenced by factors including: modelling of brittle and ductile behaviour of components, adopted fatigue damage thresholds and whether the system is able to identify applied loads. While the SHM system design is not explicitly considered in ref. [40], the study presents the possibility of defining new cost functions that are based on the value of information provided by the system, directly informing decision making.

5.3. Robustness to Benign Effects

The systems of interest are typically exposed to significant operational and environmental effects, and it is well established that the features that are typically adopted for SHM purposes also demonstrate significant sensitivity to these benign effects. Among the most highly-cited illustrations of the impact of temperature effects is the Z-24 Bridge dataset gathered by Peeters and De Roeck [41]. In addition to environmental effects, load variation is a particular concern for structures that exhibit excitation-dependent nonlinear behaviour. Accounting for benign effects is an area of active research in its own right with concepts, such as latent variable models [42] and cointegration [43] having been demonstrated to work to great effect. Nonetheless, residual effects remain and should ideally be considered during SHM system design.

Li et al. [44] propose a load-dependent sensor placement method that is based on the EI algorithm. The adaptation made is to develop an objective function that minimises the Euclidean distance between a modal estimator based on a selected sensor subset and an ideal estimator that uses all available measurement locations in a manner similar to that proposed in ref. [26]. The considered load variation refers to the location of excitation, with the magnitude of loading not varied. A case study on a six-storey truss structure with a distribution comprising six sensors being selected from 12 candidate locations is used to demonstrate that the location at which the load is applied has a marked influence on the selected sensor layout. While the robustness of the sensor network to variations in load location was not explicitly considered, this would appear to be a straightforward progression from the work presented.

5.4. Robustness to Modelling and Prediction Errors

The aforementioned study by Kammer [8] presented an early effort to assess the effect of model error on sensor placement outcomes. In recent years, this has been developed much further with the aim of accounting for errors in the underlying model used for SHM system design. Papadimitriou and Lombaert [45] consider the effect of spatially-correlated prediction errors on sensor placement outcomes. Spatially-correlated errors are considered to be important, as they influence the degree of redundancy observed in the information content of adjacent sensors. A sequential sensor placement algorithm is adopted for optimisation, and it is demonstrated that algorithms of this type remained both accurate and computationally efficient in the face of spatially-correlated prediction errors.

Castro-Triguero et al. [46] examined the influence of parametric modelling uncertainties on optimal sensor placement for the purposes of modal analysis. The effect of uncertainty is demonstrated for four SPO approaches applied to a truss bridge structure on the basis of FE model predictions. A combination of material (Young's modulus, density) and geometric (cross-sectional dimension) model parameters were varied, with uncertainties being propagated through to model predictions via Monte Carlo sampling. In addition, noise on measured outputs was simulated. The considered numerical case study illustrated that the effect of parametric uncertainties on the returned sensor configuration was significant, although some "key" sensor locations frequently recurred within the selected sensor sets.

5.5. Robustness to Sensor and Sensor Network Failure

The work presented in ref. [31] is extended in ref. [47] to cover fail-safe sensor placements. The focus of the study is the selection of sensor sets whose subsets also offer a low rate of detection error (equivalently a high probability of correct classification P_{corr}). A fail-safe fitness function is introduced that operates as follows: given a proposed mother distribution comprising N sensors, all $N-1$ child distributions are generated and their classification performance evaluated. The value adopted for the performance of the mother set is the worst P_{corr} performance of its child distributions.

The approach is illustrated for the case of impact magnitude estimation on a composite plate using time-varying strain data gathered from a network of 17 piezoceramic sensors. A sensor subset size of three was selected for illustration, making it feasible to compare the results of a GA to the optimal outcome provided by an exhaustive search. The key observation made is that, for the particular case studied, there was a trade off between achieving an optimal mother distribution, and one that is robust to the failure of one sensor. A further observation made is that the optimal sensor distribution returned by the GA displayed marked sensitivity to statistical variations in the probability of detection error, highlighting the value in pursuing robust optimisation methods. While demonstrated for strain data and the case that only a single sensor fails, the proposed fail-safe approach was deemed sufficient generally for extension to other SHM problems to be straightforward.

Kullaa [48] presented a study on methods for distinguishing between structural damage, EOVs and sensor faults. The approach is based on the fact that sensor faults are local, while structural damage and EOVs are global. EOVs are dealt with by including these effects within the training set, in a manner consistent with the supervised learning approaches discussed in section 5.3. The study is structured such that each sensor yields a single test statistic. A Gaussian Process (GP) regression model is trained across all sensor locations such that the output of individual sensors may be estimated using the output from all others. A hypothesis test (in this case the generalised likelihood ratio test (GLRT)) is then applied to identify whether the measured output of a given sensor location deviates significantly from that predicted by the GP, enabling sensor faults to be identified. While sensor optimisation is not explicitly considered, the work presents a basis for building networks that are robust to sensor failure.

More recently, ref. [49] considers fault-tolerant wireless sensor network (WSN) design for SHM. The adoption of a wireless network architecture is an attractive development for SHM systems, offering simplified deployment that obviates the need for extensive cabling. However, wireless architectures introduce an additional set of failure modes (for example, communication errors and unstable connectivity) in addition to failure of the sensor itself. They also introduce an additional energy constraint that is not considered for wired systems. The objective of the work presented in ref. [49] is to maintain a given level of fault tolerance through robustly-optimal system design. This is achieved by implementing an algorithm that identifies optimal repairing points within sensor clusters and places backup sensors at those locations. In a further adaptation, the adopted SHM algorithm makes use of decentralised computing, with the objective that the likelihood of the network remaining connected in the event of system fault is maximised.

Finally, Yi et al. [50] present a survey of recent literature in the field of sensor validation methodologies for structural health monitoring.

5.6. Scanning Laser Vibrometry

Two primary sources of candidate sensor location information have been proposed in the literature: the predictions of a (typically FE) model; and experimentally-measured data from an extensive sensor array. The advent and increasing availability of Scanning Laser Vibrometry (SLV) presents the possibility of achieving full field vibration data acquisition experimentally, removing any issues that may arise through test-model bias. The fact this may be done in a non-contacting fashion, removing structural effects such as mass loading during acquisition of candidate set data, is a further boon. Staszewski et al. [51] demonstrated the potential of scanning laser vibrometry for Lamb wave sensing. More recently Marks et al. [52] demonstrated optimisation of sensor locations for Lamb wave based damage detection using a GA allied to SLV data. However, there remain comparatively few studies that explore the use of full-field measurement techniques as a source of candidate set data.

6. Concluding Comments

This paper presents a survey and comparison of SPO approaches applicable to SHM task. The need for sensor placement optimisation techniques specific to SHM has long been recognised and promising approaches have emerged. A great number of techniques developed for other structural dynamics applications (modal analysis, system identification, and model calibration) are also of use to the SHM practitioner. Future methods are likely to be specific to the SHM methodology employed and should be developed alongside them. The central focus of this review was to highlight emerging trends specific to SHM system development, and these are briefly summarised below.

While there has been a great deal of focus on optimisation algorithms, it is the authors' opinion that as the optimisation cost is incurred at the design stage rather than online, the costs that are associated with this step are arguably of minor importance in comparison to the overall cost of developing, implementing, and maintaining an SHM system. In terms of the development of new optimisation algorithms, two recommendations present themselves. The first is that prior to presenting the outcomes of optimising an SHM system design, it is generally desirable that the best possible performance of the system (as defined in Section 4.3) be evaluated to demonstrate whether an acceptable level of performance is achievable. Secondly, it is desirable that the performance of an "optimal" sensor distribution returned via a newly-proposed algorithm should be compared with the results of an exhaustive search wherever feasible. This is admittedly limited to cases with low numbers of sensors in the candidate and measurement sets.

There remains scope for being more ambitious in the damage-sensitive features adopted. This ambition has two strands. The first is that there remains a focus in large parts of the literature on modal features alone, although a smaller number of the reviewed studies considered other options based on strain field measurements and time of flight of wave packets for impact detection, for example. Methods appropriate to general feature sets including multiple sensor modalities—the domain of data fusion—remain an emerging topic. The second area for greater ambition is in incorporating feature selection within the sensor placement task i.e the selection of maximally sensitive feature sets within the data gathered from a given sensor distribution). A common approach in the literature is to set the features to be returned by a sensor location at an early stage (e.g., a vector of modeshape elements). While removing such constraints may inevitably, and perhaps dramatically, increase the computational expense associated with the optimisation task, it may extend the upper bound on "best possible" performance.

As highlighted in Section 5.4, there have been some efforts towards developing SPO methods that are robust to model uncertainty and experimental variability, but there remains scope to go much further with this in order to increase confidence in the SHM system design process. There also remains a question mark over how to overcome the difficulty of sourcing damaged state data with which to train SHM classifiers. The sources of data used for SHM system optimisation fall broadly into two categories: the predictions of a numerical model of a given structure, or experimental data gathered from a structure in situ. The former allows for consideration of the effects of damage, but legitimate questions may be posed around the required accuracy of damaged-state model predictions. The latter avoids the issue of test-model bias but will typically be limited by the lack of system-level damaged-state data (although advances are being made in the use of Population-Based methods that attempt to address this [53]) The overwhelming majority of studies reviewed focus purely on development and evaluation using numerical data; or alternatively demonstrated these steps using purely experimental data. It is noted that there remains a distinct shortage of work concentrating on the practical, experimental evaluation of systems developed using *a priori* physics-based modelling on realistic structures. Finally, there appears to be broad scope for the further development of methods that focus on optimisation of SHM decision outcomes, as opposed to classification outcomes alone.

Author Contributions: R.B.: Conceptualization, writing—original draft preparation; K.W.: Conceptualization, writing—review and editing. All authors have read and agreed to the published version of the manuscript.

Funding: This work was supported by the Engineering and Physical Sciences Research Council [Grant numbers EP/J013714/1 and EP/N010884/1].

Conflicts of Interest: The authors declare no conflict of interest.

Abbreviations

The following abbreviations are used in this manuscript:

SHM	Structural health monitoring
SPO	Sensor placement optimisation
DOFs	Degrees of freedom
FEA	Finite element analysis
EO	Evolutionary optimisation
FIM	Fisher information matrix
EOV	Experimental and operational variation
MAC	Modal assurance criterion
NMSE	Normalised mean squared error
GA	Genetic algorithm
EI	Effective independence
SSP	Sequential sensor placement
ROC	Receiver operating characteristic

References

1. Barthorpe, R.J.; Worden, K. Sensor Placement Optimization. In *Encyclopedia of Structural Health Monitoring*; Boller, C., Chang, F., Fujino, Y., Eds.; John Wiley & Sons: Chichester, UK, 2009. [CrossRef]
2. Yi, T.H.; Li, H.N. Methodology Developments in Sensor Placement for Health Monitoring of Civil Infrastructures. *Int. J. Distrib. Sens. Netw.* **2012**, *8*. [CrossRef]
3. Ostachowicz, W.; Soman, R.; Malinowski, P. Optimization of sensor placement for structural health monitoring: A review. *Struct. Health Monit.* **2019**, *18*, 963–988. [CrossRef]
4. Papadimitriou, C. Optimal sensor placement methodology for parametric identification of structural systems. *J. Sound Vib.* **2004**, *278*, 923–947. [CrossRef]
5. Udwadia, F. Methodology for Optimum Sensor Locations for Parameter Identification in Dynamic Systems. *ASCE J. Eng. Mech.* **1994**, *120*. [CrossRef]
6. Kammer, D. Sensor placement for on-orbit modal identification and correlation of large space structures. *J. Guid. Control. Dyn.* **1991**, *14*, 251–259. [CrossRef]
7. Shah, P.; Udwadia, F. A Methodology for Optimal Sensor Locations for Identification of Dynamic Systems. *J. Appl. Mech.* **1978**, *45*, 188–196. [CrossRef]
8. Kammer, D. Effect of model error on sensor placement for on-orbit modal identification of large space structures. *J. Guid. Control. Dyn.* **1992**, *15*, 334–341. [CrossRef]
9. Kammer, D. Effects of noise on sensor placement for on-orbit modal identification of large space structures. *J. Dyn. Syst. Meas. Control. Trans. ASME* **1992**, *114*, 436–443. [CrossRef]
10. Yao, L.; Sethares, W.; Kammer, D. Sensor placement for on-orbit modal identification via a genetic algorithm. *AIAA J.* **1993**, *31*, 1922–1928. [CrossRef]
11. Kammer, D. Sensor set expansion for modal vibration testing. *Mech. Syst. Signal Process.* **2005**, *19*, 700–713. [CrossRef]
12. Penny, J.; Friswell, M.; Garvey, S. Automatic choice of measurement locations for dynamic testing. *AIAA J.* **1994**, *32*, 407–414. [CrossRef]
13. Salama, M.; Rose, T.; Garba, J. Optimal placement of excitations and sensors for verification of large dynamical systems. In Proceedings of the 28th Structures, Structural Dynamics and Materials Conference, Monterey, CA, USA, 6–8 April 1987. [CrossRef]
14. Li, D.S.; Li, H.N.; Fritzen, C.P. The connection between effective independence and modal kinetic energy methods for sensor placement. *J. Sound Vib.* **2007**, *305*, 945–955. [CrossRef]
15. Imamovic, N.; Ewins, D. Optimization of excitation DOF selection for modal tests. In Proceedings of the 15th International Modal Analysis Conference, Orlando, FL, USA, 3–6 February 1997.
16. Imamovic, N. Validation of Large Structural Dynamics Models Using Modal Test Data. Ph.D. Thesis, Imperial College London, London, UK, 1998.

17. Larson, C.B.; Zimmerman, D.C.; Marek, E.L. A comparison of modal test planning techniques: Excitation and sensor placement using the NASA 8-bay truss. In Proceedings of the 12th International Modal Analysis Conference, Honolulu, HI, USA, 31 January–3 February 1994; pp. 205–211.
18. Shi, Z.; Law, S.; Zhang, L. Optimum sensor placement for structural damage detection. *J. Eng. Mech.* **2000**, *126*, 1173–1179. [CrossRef]
19. Guo, H.; Zhang, L.; Zhang, L.; Zhou, J. Optimal placement of sensors for structural health monitoring using improved genetic algorithms. *Smart Mater. Struct.* **2004**, *13*, 528–534. [CrossRef]
20. Hemez, F.; Farhat, C. An energy based optimum sensor placement criterion and its application to structural damage detection. In Proceedings of the SPIE International Society for Optical Engineering, Bergen, Norway, 13–15 June 1994; pp. 1568–1568.
21. Said, W.M.; Staszewski, W.J. Optimal sensor location for damage detection using mutual information. In Proceedings of the 11 th International Conference on Adaptive Structures and Technologies (ICAST), Nagoya, Japan, 23–26 October 2000; pp. 428–435.
22. Trendafilova, I.; Heylen, W.; Van Brussel, H. Measurement point selection in damage detection using the mutual information concept. *Smart Mater. Struct.* **2001**, *10*, 528–533. [CrossRef]
23. Papadimitriou, C.; Beck, J.; Au, S.K. Entropy-based optimal sensor location for structural model updating. *Jvc/J. Vib. Control.* **2000**, *6*, 781–800. [CrossRef]
24. Tharwat, A. Classification assessment methods. *Appl. Comput. Inform.* **2018**. [CrossRef]
25. Fawcett, T. An introduction to ROC analysis. *Pattern Recognit. Lett.* **2006**, *27*, 861–874. [CrossRef]
26. Worden, K.; Burrows, A. Optimal sensor placement for fault detection. *Eng. Struct.* **2001**, *23*, 885–901. [CrossRef]
27. Papadimitriou, C. Pareto optimal sensor locations for structural identification. *Comput. Methods Appl. Mech. Eng.* **2005**, *194*, 1655–1673. [CrossRef]
28. Stephan, C. Sensor placement for modal identification. *Mech. Syst. Signal Process.* **2012**, *27*, 461–470. [CrossRef]
29. Stabb, M.; Blelloch, P. A genetic algorithm for optimally selecting accelerometer locations. In Proceedings of the 13th International Modal Analysis Conference, Nashville, TN, USA, 13–16 February 1995; Volume 2460, p. 1530.
30. Worden, K.; Burrows, A.; Tomlinson, G. A combined neural and genetic approach to sensor placement. In Proceedings of the 15th International Modal Analysis Conference, Nashville, TN, USA, 12–15 February 1995; pp. 1727–1736.
31. Worden, K.; Staszewski, W. Impact location and quantification on a composite panel using neural networks and a genetic algorithm. *Strain* **2000**, *36*, 61–70. [CrossRef]
32. Coverley, P.; Staszewski, W. Impact damage location in composite structures using optimized sensor triangulation procedure. *Smart Mater. Struct.* **2003**, *12*, 795. [CrossRef]
33. De Stefano, M.; Gherlone, M.; Mattone, M.; Di Sciuva, M.; Worden, K. Optimum sensor placement for impact location using trilateration. *Strain* **2015**, *51*, 89–100. [CrossRef]
34. Overton, G.; Worden, K. Sensor optimisation using an ant colony methapor. *Strain* **2004**, *40*, 59–65. [CrossRef]
35. Scott, M.; Worden, K. A Bee Swarm Algorithm for Optimising Sensor Distributions for Impact Detection on a Composite Panel. *Strain* **2015**, *51*, 147–155. [CrossRef]
36. Beal, J.M.; Shukla, A.; Brezhneva, O.A.; Abramson, M.A. Optimal sensor placement for enhancing sensitivity to change in stiffness for structural health monitoring. *Optim. Eng.* **2007**, *9*, 119–142. [CrossRef]
37. Wolpert, D.H.; Macready, W.G. No free lunch theorems for optimization. *IEEE Trans. Evol. Comput.* **1997**, *1*, 67–82. [CrossRef]
38. Cawley, P. Structural health monitoring: Closing the gap between research and industrial deployment. *Struct. Health Monit.* **2018**, *17*, 1225–1244. [CrossRef]
39. Flynn, E.; Todd, M. A Bayesian approach to optimal sensor placement for structural health monitoring with application to active sensing. *Mech. Syst. Signal Process.* **2010**, *24*, 891–903. [CrossRef]
40. Thöns, S. On the Value of Monitoring Information for the Structural Integrity and Risk Management. *Comput. Civ. Infrastruct. Eng.* **2018**, *33*, 79–94. [CrossRef]
41. Peeters, B.; De Roeck, G. One-year monitoring of the Z24-Bridge: Environmental effects versus damage events. *Earthq. Eng. Struct. Dyn.* **2001**, *30*, 149–171. [CrossRef]

42. Kullaa, J. Eliminating Environmental or Operational Influences in Structural Health Monitoring using the Missing Data Analysis. *J. Intell. Mater. Syst. Struct.* **2009**, *20*, 1381–1390. [CrossRef]
43. Cross, E.J.; Worden, K.; Chen, Q. Cointegration: A novel approach for the removal of environmental trends in structural health monitoring data. *Proc. R. Soc. A* **2011**, *467*, 2712–2732. [CrossRef]
44. Li, D.S.; Li, H.N.; Fritzen, C.P. Load dependent sensor placement method: Theory and experimental validation. *Mech. Syst. Signal Process.* **2012**, *31*, 217–227. [CrossRef]
45. Papadimitriou, C.; Lombaert, G. The effect of prediction error correlation on optimal sensor placement in structural dynamics. *Mech. Syst. Signal Process.* **2012**, *28*, 105–127. [CrossRef]
46. Castro-Triguero, R.; Murugan, S.; Gallego, R.; Friswell, M.I. Robustness of optimal sensor placement under parametric uncertainty. *Mech. Syst. Signal Process.* **2013**, *41*, 268–287. [CrossRef]
47. Side, S.; Staszewski, W.; Wardle, R.; Worden, K. Fail-safe sensor distributions for damage detection. In Proceedings of the DAMAS'97 International Workshop on Damage Assessment using Advanced Signal Processing Procedures, Sheffield, UK, 30 June–2 July 1997; pp. 135–146.
48. Kullaa, J. Distinguishing between sensor fault, structural damage, and environmental or operational effects in structural health monitoring. *Mech. Syst. Signal Process.* **2011**, *25*, 2976–2989. [CrossRef]
49. Bhuiyan, M.; Wang, G.; Cao, J.; Wu, J. Deploying Wireless Sensor Networks with Fault-Tolerance for Structural Health Monitoring. *IEEE Trans. Comput.* **2015**, *64*, 382–395. [CrossRef]
50. Yi, T.H.; Huang, H.B.; Li, H.N. Development of sensor validation methodologies for structural health monitoring: A comprehensive review. *Measurement* **2017**, *109*, 200–214. [CrossRef]
51. Staszewski, W.; Lee, B.; Mallet, L.; Scarpa, F. Structural health monitoring using scanning laser vibrometry: I. Lamb wave sensing. *Smart Mater. Struct.* **2004**, *13*, 251. [CrossRef]
52. Marks, R.; Clarke, A.; Featherston, C.A.; Pullin, R. Optimization of acousto-ultrasonic sensor networks using genetic algorithms based on experimental and numerical data sets. *Int. J. Distrib. Sens. Netw.* **2017**, *13*. [CrossRef]
53. Gardner, P.; Liu, X.; Worden, K. On the application of domain adaptation in structural health monitoring. *Mech. Syst. Signal Process.* **2020**, *138*, 106550. [CrossRef]

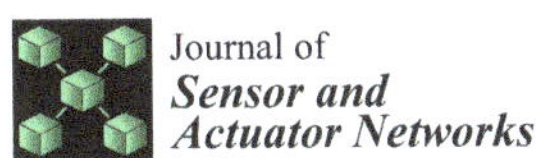
Journal of
Sensor and Actuator Networks

Article

The Design and Calibration of Instrumented Particles for Assessing Water Infrastructure Hazards

Khaldoon Al-Obaidi, Yi Xu and Manousos Valyrakis *

Water Engineering Lab, School of Engineering, University of Glasgow, Glasgow G12 8QQ, UK; 2372435A@student.gla.ac.uk (K.A.-O.); 2421380X@student.gla.ac.uk (Y.X.)
* Correspondence: Manousos.Valyrakis@glasgow.ac.uk; Tel.: +44-(0)-141-330-5209

Received: 9 May 2020; Accepted: 27 July 2020; Published: 30 July 2020

Abstract: The highly dynamical entrainment and transport processes of solids due to geophysical flows is a major challenge studied by water infrastructure engineers and geoscientists alike. A miniaturised instrumented particle that can provide a direct, non-intrusive, low-cost and accessible method compared to traditional approaches for the assessment of coarse sediment particle entrainment is developed, calibrated and tested. The instrumented particle presented here is fitted with inertial microelectromechanical sensors (MEMSs), such as a triaxial accelerometer, a magnetometer and angular displacement sensors, which enable the recording of the particle's three-dimensional displacement. The sensor logs nine-axis data at a configurable rate of 200–1000 Hz and has a standard mode of deployment time of at least one hour. The data can be obtained and safely stored in an internal memory unit and are downloadable to a PC in an accessible manner and in a usable human-readable state. A plethora of improved design specifications have been implemented herein, including increased frequency, range and resolution of acceleration and gyroscopic sensing. Improvements in terms of power consumption, in comparison to previous designs, ensure longer periods of data logging. The embedded sensors are calibrated using simple physical motions to validate their operation. The uncertainties in the experiments and the sensors' readings are quantified and an appropriate filter is used for inertial sensor fusion and noise reduction. The instrumented particle is tested under well-controlled lab conditions, where the beginning of the destabilisation of a bed surface in an open channel flow, is showcased. This is demonstrative of the potential that specifically designed and appropriately calibrated instrumented particles have in assessing the initiation and occurrence of water infrastructure hazards.

Keywords: inertial sensor fusion; instrumented particle; MEMS; sediment entrainment; sensor calibration; frequency of entrainment

---

1. Introduction

The surface of a planet is shaped by geomorphic processes, the majority of which are driven by a range of geophysical flows, including turbulent fluvial, aeolian, snow/ice and lava flows. Of special interest for geomorphologists and civil and environmental engineers alike are turbulent flows in rivers, canals, estuaries and coasts that can set into entrainment and transport coarse sediment particles, which have the potential to destabilise critical infrastructure found on their way and result in significant geophysical hazards. Coarse particle entrainment in turbulent flows has been shown to occur due to the existence of strong energetic events that can result in its removal from its resting location [1]. Such transport processes are considered to be the governing mechanism for the failure of built infrastructure, such as bridge piers and abutments, as well as the destabilisation of riverbanks and embankments, rendering imperative the monitoring of the starting phase of these processes to reduce risk and improve resilience [2–4]. These processes depend on the particle and flow parameters, in addition to any specific water infrastructure designs present at the studied area. Despite recent advances in the prediction of the turbulent flow conditions that may result in entraining a particle [5–7], such criteria have yet to find widespread practical application in engineering. Instead, relatively expensive, indirect and highly inaccurate methods for detecting mean flow parameters to determine the possibility of sediment entrainment are still employed. In general, such studies use expensive acoustic Doppler velocimetry [8], laser Doppler velocimetry [9] or water level stations, along with discharge hydrographs, requiring regular or reactive visits to remote and harsh sites [10]. Recent technological advancements make it possible for sediment entrainment to be assessed directly, using novel tools, instead of traditionally monitoring surrogate flow metrics [2,11]. Using sensors in the field of riverbed sediment transport has been explored by multiple researchers [12–19]. However, none of the work presented in literature has used such smart sensors to establish a link between logged readings and sediment entrainment based on derived performance indicators. This work introduces the methodology to derive performance indicators, like frequency of entrainment, based on the particle's fused data, which can give information on the risk of earth surface destabilisation or the risk of scour development. Additionally, the presented instrumented particle in this work has a plethora of improved design specifications compared to other tools that exist in the literature, including a decreased particle, increased frequency and range of sensor readings, reduced error after calibration and fusion and improved battery life for longer deployment periods.

In this study, a miniaturised instrumented particle has been designed and developed to help monitor the above entrainment processes. The presented design capitalises on improvements in hardware, software and casing designs, stemming from the many years' experience in further developing a number of earlier prototypes [1–3], to allow improved and higher resolution particle motion detection. Additionally, this work presents the calibration and testing of the instrumented particle entrainments due to the action of near bed turbulent flow events at one of the open channel flumes of the Water Engineering Lab at the University of Glasgow, and the inertial sensor fusion of logged readings to achieve uncertainty reduction in the gathered dynamical data. Finally, the research herein suggests using MEMS instrumented particles laying on the riverbed surface for monitoring its destabilisation potential, using the frequency of entrainment as a performance indicator (Supplementary Materials).

To achieve this task, a 40 mm diameter data-logging device was designed, encapsulated in a 3D-printed plastic spherical shell. The device was designed for use in a laboratory setting (such as in water and sediment recirculating flumes), as well for as real world testing, on the bed surface of streams and rivers, for monitoring the potential for geomorphic work. Initial designs of the instrumented particle ranged from 125 to 75 mm in size [1–3]. However, riverbeds are characterised by a wider range of bed surface material and, in many cases, the characteristic size of sediment particles may be smaller than that of the previous designs. The decrease in size of the proposed instrumented particle was necessary in order to allow for widening its range of application by enabling the matching of the particle characteristics and assessing the possibility of sediment entrainment for certain flow conditions more effectively [1–3].

In addition to the reduced size, additional improvements in design specifications included a higher frequency and range of acceleration and gyroscopic sensing, and inertial sensor fusion for reducing the uncertainty in tracking the motion of the particle. Hardware modifications, resulting in lower power consumption, allowed for achieving longer logging periods. The modular design can be extended to embed a range of additional sensors, such as magnetometers, water temperature sensors, static and dynamic pressure transducers (for flow depth monitoring and dynamic flow motions) and photocell sensors (for ecological applications). Even though those sensors have a promising potential for future deployment in a wider range of ecohydraulic and geoscience applications, focus was given to the fusion of triaxial acceleration and gyroscope readings with the magnetometer.

The data rate of the sensors ranged from 200 Hz up to 1000 Hz. A one-hour run time was also required for robust field deployment and recording the instrumented particle's inertial data. Reducing the size of the instrumented particle was a challenging task, as it required a high level of integration and optimisation of the internal components, both in terms of physical space (circuit design) and power consumption. The physical restrictions imply that function was improved, while the instrumented particle's volume was reduced by more than nine times compared to previous designs, meaning that all subsystems were integrated completely into a truly bespoke solution, as discussed in the following sections.

The rest of this paper consists of five sections that present: the design considerations, the calibration, the flume testing, the results and discussion and the conclusions and future work. Section 2 introduces the main considerations for the design of the instrumented particle for the applications discussed previously. Section 3 describes the calibration process to estimate the uncertainties in the sensor's readings and the inertial sensor fusion filter used to reduce these uncertainties. Section 4 presents the flume testing of the instrumented particle by performing three experiments for a certain flowrate range that represents the near-threshold conditions. Section 5 introduces the results and demonstrates how the sensor's logged readings could be used to assess the probability of sediment entrainment. Finally, Section 5 presents the conclusions and future work.

2. Design Considerations

The main considerations for the design of the instrumented particle are: the size and availability of the sensor, the energy supply, the flash storage, the microcontroller, the inertial sensor, weight balance and data transfer [20]. The target design requirements for the instrumented particle and the sensor are described in Table 1 below:

Table 1. Main design considerations.

Size of instrumented particle	≤4 cm in diameter
Size of sensor	≤4 cm in diameter
Time of operation	>1 h
Logging frequency range	>200 Hz
Maximum acceleration	>10 g
Maximum rotational velocity	>1000°/s
Data transfer rate from sensor to PC	As fast and practical as possible
Energy supply	Cost effective and practical
Flash storage	Predictable access, erase and write times. Logging speed of >5 KB/s
Microcontroller	Simplicity, strong documentation and availability of test data
Inertial sensor	Six or more axes (triaxial accelerometer and triaxial gyroscope) Communication over I2C or SPI

Based on the criteria specified in Table 1, the final decision is made to select Invensense MPU-9250 inertial sensor (Invensense Inc., San Jose, CA, USA). It supports three-axis acceleration, three-axis rotational velocity, and a three-axis magnetometer, as well as a well-renowned internal Digital Motion Processor, supporting up to 16 g of measurable acceleration, up to 2000°/s of measurable rotational velocity and sensor output frequency well in excess of 200 Hz [21]. As for the energy supply, rechargeable coin cells (Varta Microbattery produced under CoinPower [22], Varta Microbattery GmbH, Ellwangen Germany) are selected due to being a cost effective and practical option compared to lithium batteries [23,24]. For the flash storage, discrete flash integrated circuits are chosen due to more predictable access, erase and write times compared to an SD card. S25FL128S (Cypress Semiconductor Corporation, San Jose, CA, USA) is chosen specifically due its low cost, its 16 MB size and high data logging speed of 1.5 MB/s, which is far faster than the required 5 KB/s (transferring roughly 16 MB of logged data for one hour of field deployment). As for the microcontroller, the Arduino Zero (Adruino, Ivrea Italy) is chosen for its higher likelihood of communicating, computing and storing data within the necessary time frame, as shown in Figure 1. The final design of the sensor is shown in Figure 2 and specific attention has been paid to centring the weight of the sensor to produce a well-balanced instrumented particle. Finally, for data transfer, Tera Term, an open source software, is used to establish the communication between the microcontroller and the inertial sensor. The software allows the user to erase data from the flash, transfer data to the SD card and start the logging process. The rate of data transfer from the flash to the SD card (even though it can be further optimised) is sufficiently fast (requiring about half of the deployment time to transfer the logged readings).

Figure 1. The microcontroller and docking station used for charging the inertial sensor and transferring the data to an SD card.

Figure 2. The 40 mm instrumented particle printed circuit board design.

3. Calibration and Uncertainty Estimation

3.1. Checking Sensor Response and Range for Vigorous Rolling Motion

Before the calibration process starts, a quick sensing check and endurance experiment is performed. This involves inserting the sensors into the spherical shell of the instrumented particle (4 cm in external diameter), fixing them properly so that they are centred at their centre of gravity and allowing them to vigorously roll on a flat surface, by hand. The resulting motion is checked by separately observing the readings of the accelerometer, gyroscope and magnetometer over the three axes, as shown in Figure 3.

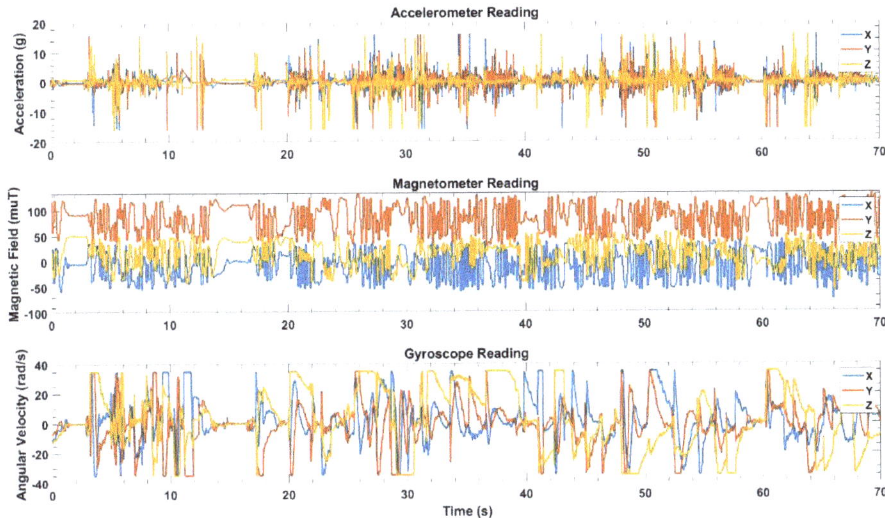

Figure 3. Accelerometer, gyroscope and magnetometer logged readings during a quick sensing check and endurance experiment that is performed by throwing the sensor to roll on the floor.

From this simple test, it is observed that the selected ranges of 16 g and 2000°/s are sufficient for recording the maximum particle accelerations and gyroscope readings. Additionally, the accelerometer recorded the gravitational acceleration (shown as a recording of 1 g) on the z-axis during the times when the sensor is static. Finally, after throwing the sensor with a high force, it remains operating and functional, which confirms that it could be used in harsh and remote water environments, such as in fluvial systems.

3.2. Calibration for Controlled Harmonic Motion

The calibration process consisted of individually calibrating the accelerometer and the gyroscope. Specifically, for the calibration of the accelerometer, a simple harmonic motion of a pendulum with the instrumented particle at its centre of gravity and a shaking table of a known acceleration range are used. The pendulum test consists of two parts, where the first part involves measuring the pendulum angle and checking the accelerometer's reading at the lowest point of the swing, and the second involves the oscillation period of the pendulum, using a stop watch and the accelerometer's readings and comparing both to the oscillation period that is estimated theoretically. As for the first step, the accelerometer data are used to determine the total acceleration of the sensor using Equation (1). The total acceleration at the beginning of the experiment is zero since the sensor is let from rest then it increases in value till it reaches a maximum at the lowest point of the motion, since the angle is the highest, as illustrated in Figure 4. The acceleration of the sensor at the lowest point is then calculated using Equation (2).

$$a = \sqrt{a_x^2 + a_y^2 + a_z^2} \tag{1}$$

where a_x, a_y and a_z are the x, y and z components of the acceleration and

$$a_{@\text{lowest point}} = g \sin\theta \tag{2}$$

where θ is the angle of the pendulum, as illustrated in Figure 4, and g is the gravitational acceleration (9.81 m/s^2).

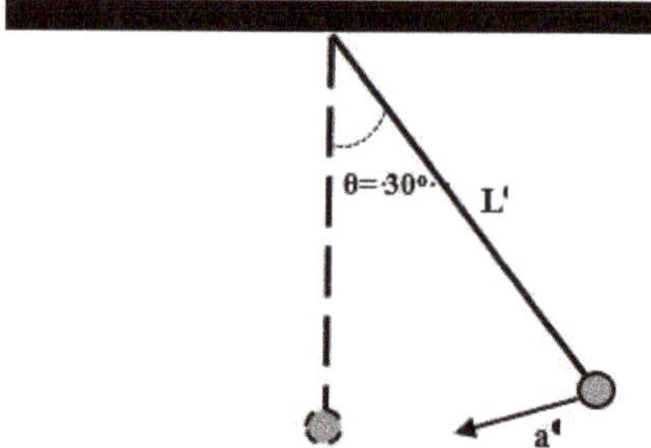

Figure 4. An illustration of a simple harmonic motion of a pendulum.

A starting pendulum angle (θ) of 30° is selected, as shown in Figure 4. The experiment is repeated 10 times and the total acceleration results at the lowest point are estimated using (1), based on the accelerometer's readings for the 10 runs, as shown in Table 2. The standard deviation for the 10 runs is 0.0073 which is a small value, indicating a small spread in the results. The standard deviation per mean is estimated to be 0.0146, indicating that the error in the experiment is small. As for the average error of the 10 runs, it is 1.08%, which incorporates both experimental measurement errors (as the particle is released by hand) and estimation errors due to the mechanical energy lost due to friction with the air, not accounted for in Equation (2). Thus, the average error in the accelerometer's readings for this experiment is deemed to be acceptable.

Table 2. The acceleration results for 10 runs of pendulum swinging, starting from a known angle.

Run Name	Highest Recorded Total Acceleration (9.81 × m/s^2)	Expected Acceleration (9.81 × m/s^2) Estimated Using Equation (2)	% Error
Run1	0.516	0.500	3.28
Run2	0.492		1.54
Run3	0.508		1.6
Run4	0.498		0.34
Run5	0.497		0.56
Run6	0.502		0.30
Run7	0.503		0.54
Run8	0.501		0.22
Run9	0.489		2.2
Run10	0.499		0.18

The second part of the experiment is to estimate the duration of a full swing for the pendulum using the accelerometer's readings. The pendulum's single full swing period recorded by the accelerometer (Figure 5) corresponds to half a period of the pendulum motion [25]. The theoretical full swing period of a simple pendulum (the time it takes to swing back-and-forth) in this experiment, estimated using Equation (3), is 1.30 s:

$$T = 2\pi\sqrt{\frac{L}{g}} \tag{3}$$

where T is the period of a simple pendulum, L is the length from the pivot point to the center of mass of the sensor (42 cm in this experiment) g is the gravitational acceleration. The experiment is repeated 10 times and the swing period results, estimated using the accelerometer's readings for the 10 runs, are shown in Table 3.

The standard deviation for the 10 periods of the 10 runs in the experiment is 0.0234, which is a small value, indicating a small spread in the results. The standard deviation per mean is estimated to be 0.0180, indicating that the error in the experiment is small. The average error in the 10 runs is 1.41%

and, considering the uncertainties in assessing the parameters in Equation (3) and the error due to the mechanical energy lost due to friction with the air, it is deemed to be acceptable.

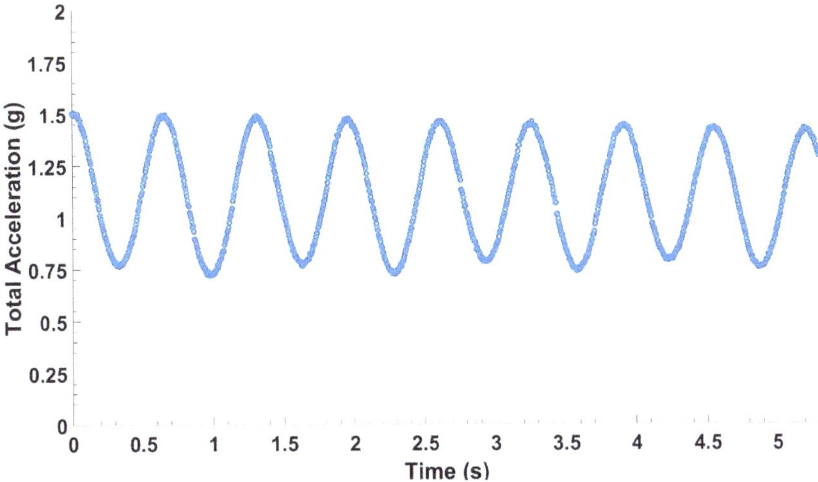

Figure 5. Total acceleration estimated using the accelerometer logged readings corresponds to half of the theoretical period estimated using Equation (3).

Table 3. The swing period results of 10 runs of pendulum swinging.

Run Name	Period (s) Estimated Using the Accelerometer's Readings	Theoretical Period (s) Estimated Using Equation (3)	% Error
Run1	1.286		1.05
Run2	1.302		0.15
Run3	1.265		2.72
Run4	1.313		1.02
Run5	1.312	1.30	0.92
Run6	1.270		2.31
Run7	1.298		0.14
Run8	1.310		0.77
Run9	1.285		1.18
Run10	1.350		3.85

3.3. Calibration on a Shaking Table

The next experiment involves attaching the sensor to a shaking table with a back-and-forth movement of a known acceleration of 6 g. The representative time series for the acceleration in the direction of movement (achieved by aligning the sensor's local coordinate system with that of the shake table) is shown in Figure 6.

The highest value recorded by the accelerometer is 6.27 g. It is clear that any errors in this experiment are not due to the sensor alone but also because the sensor is not rigidly fixed on the shaking table, thus the full- and without-time delay and the transfer of motion cannot be practically achieved. The error for this experiment is 4.5%, which is acceptable, considering that accelerations are typically far below this range for applications in water environments and for this size and submerged density of the instrumented particle.

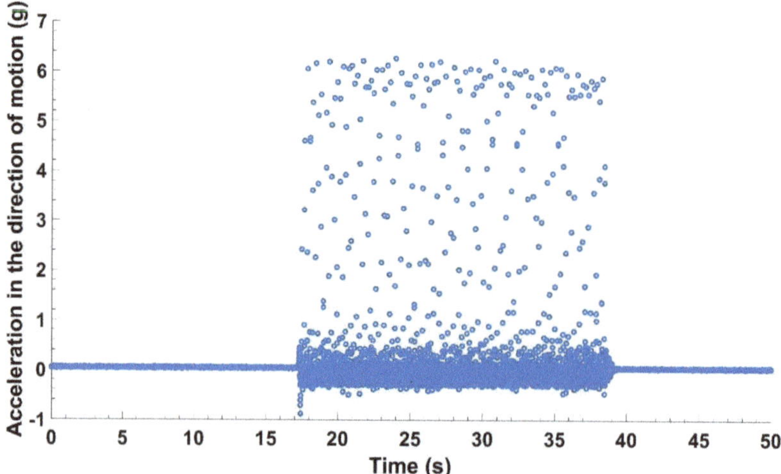

Figure 6. The accelerometer logged readings in the direction of motion of the shaking table.

3.4. Calibration of Gyroscope for Pure Rolling along an Incline

As for the gyroscope calibration, it consists of attaching the sensor to the centre of a hollow cylinder and letting it roll down an inclined plane of 5 cm in height, starting from rest, as illustrated in Figure 7.

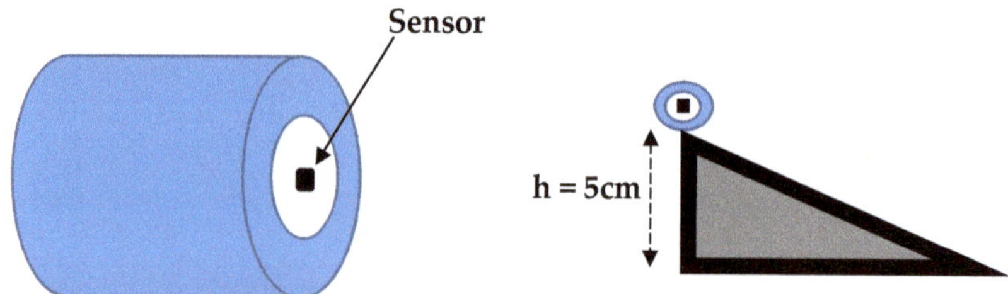

Figure 7. An illustration of rolling the sensor attached to the centre of a hollow cylinder down an inclined plane of 5 cm in height.

To ensure the repeatability of the experiment and minimise the experimental error, a thread is attached to the start location of the roll to ensure each run starts from the same position and a hard surface is placed at the end of the plane to keep the inclined plane in the same location. For a pure roll down an inclined plane, the conservation of energy can be used to estimate the angular velocity down the inclined plane since the potential lost by the drop in height should be equal to the gain in kinetic energy down the plane, as shown in Equation (4), assuming zero friction between the cylinder and plane surface:

$$\mathrm{mgh} = \frac{1}{2}\mathrm{mv}^2 + \frac{1}{2}\mathrm{I}\omega^2 \tag{4}$$

where m is the mass, h is the height of the inclined plane, v is the linear velocity, I is the moment of inertia and ω is the angular velocity. The moment of inertia of the hollow cylinder depends on its inner and outer radii, as shown in Equation (5):

$$\mathrm{I} = \frac{1}{2}\mathrm{m}\left(\mathrm{a}^2 + \mathrm{b}^2\right) \tag{5}$$

where a is the inner radius and b is the outer radius of cylinder.

Substituting Equation (4) in Equation (5) and simplifying yields:

$$\omega = \sqrt{\frac{2gh}{b^2 + \frac{1}{2}\left(b^2 + a^2\right)}} \tag{6}$$

with inner and outer radii of the hollow cylinder of 40 and 55 mm, respectively, and a height of the inclined plane of 5 cm, the angular velocity theoretically should be 13.55 rad/s. The angular velocity is calculated using the sensor's gyroscope data as $\omega = \sqrt{\omega_x^2 + \omega_y^2 + \omega_z^2}$. The maximum recorded angular velocity results by the gyroscope for the 10 runs are shown in Table 4, which corresponds to the angular velocity at the bottom of the inclined plane, as shown in Figure 8. As for the sudden drop in the total angular velocity after it reached a maximum, this is due to the hard surface that is set at the bottom of the inclined plane.

Table 4. The total angular velocity results of 10 runs of pure rolling down an incline.

Run Name	Total Angular Velocity (rad/s) Estimated Using the Gyroscope's Readings	Theoretical Total Angular Velocity (rad/s) Estimated Using Equation (6)	% Error
Run1	13.40	13.55	1.11
Run2	13.92		2.73
Run3	13.32		1.70
Run4	13.10		3.32
Run5	13.65		0.74
Run6	13.42		0.96
Run7	14.30		5.54
Run8	13.69		1.03
Run9	13.45		0.74
Run10	13.25		2.21

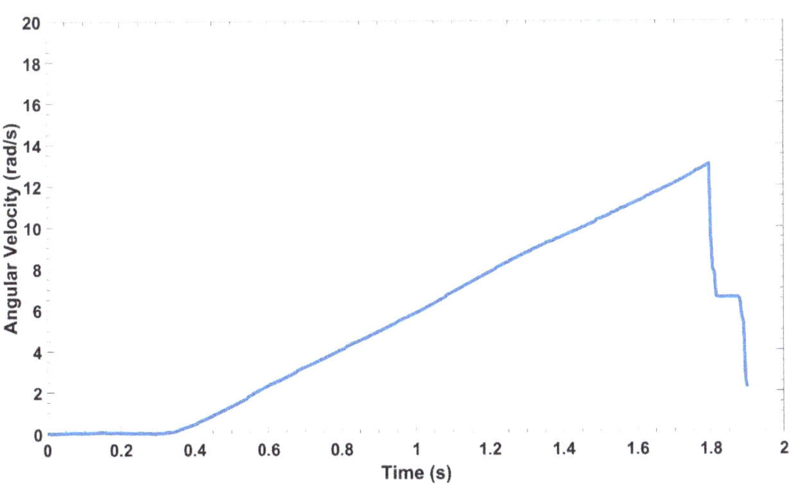

Figure 8. Total angular velocity recorded by the sensor during a roll from the top to the bottom of the inclined plane.

The standard deviation for the 10 runs in the experiment is 0.335, which is a small value, indicating a small spread in the results. The standard deviation per mean is estimated to be 0.0247, indicating that the error in the experiment is small. The average error in the 10 runs is 2.01%, which is deemed acceptable, considering the fact that the measurements are made by hand and the theoretical estimation using Equation (6) is based on the assumption that there is no sliding motion between the cylinder and the inclined plane. Additionally, the data used for the calculations are the raw data before the inertial sensor fusion, i.e., before using the accelerometer data to correct the gyroscope data since the gyroscope's data are less reliable due to drifting [26,27].

3.5. Data Fusion

The calibration of the accelerometer includes two different experiments: simple harmonic motion and using a shaking table, with average error values of 1.08% and 4.5%, respectively. On the other hand, the calibration of the gyroscope includes pure rolls down an inclined plane experiment, with an error of 2.01%. The range of errors arising from the sensor's readings and from the experimental measurements are quantified in this calibration process. The next step involves inverse uncertainty quantification by parameter calibration and data fusion. Specifically, this involves changing the parameters included in the input of a filter to be used for sensor data fusion to estimate the final corrected acceleration, angular velocity and orientation, based on the experimental results and the known information about the sensors. The filter itself uses nine-axis Kalman [28] filter structures with inputs of the expected accelerometer, gyroscope and magnetometer noise. It is decided to use the average of the mean square error of the pendulum period estimation and acceleration estimation of the pendulum swinging experiments as the accelerometer noise and the mean square error of pure rolls down an incline experiment as the gyroscope noise for the input of the inertial sensor fusion filter. Specifically, accelerometer noise of 0.0029608 $(m/s^2)^2$ and gyroscope noise of 0.11238 $(rad/s)^2$ are used as the inputs for the inertial sensor fusion filter.

4. Testing

4.1. Flume Setup and Test Section

In order to test the instrumented particle, a well-controlled laboratory experiment is performed in an 8 m long water recirculating flume. This 0.9 m wide water recirculating open channel is able to carry flows of up to 0.4 m deep, while water is provided with a maximum capacity of 0.2 m^3/s, controlled by a torque inverter though which the operating frequency of the pump can be adjusted. In order to artificially create the intended flow depths at reasonable flow velocities, an adjustable tailgate is located at the outlet, while the bed surface slope can be adjusted. In order to achieve an adequate hydraulic roughness, the bed surface is paved with a few layers of water-worked uniformly sized fine gravel with a median size of about d_{50} = 15–25 mm.

The test section (has a length of 1 m) is located in the downstream section of the flume, approximately 5.25 m downstream from the inlet and 1.75 m upstream from the outlet (Figure 9a), in order to ensure that the hydraulically rough turbulent flow is fully developed. The test section is positioned along the centreline of the flume, 45cm from either of the flume's side glass walls. An illustration of the flume with the instrumented particle and local bed microtopography is shown in Figure 9b, demonstrating the instrumented particle's initial and displaced position.

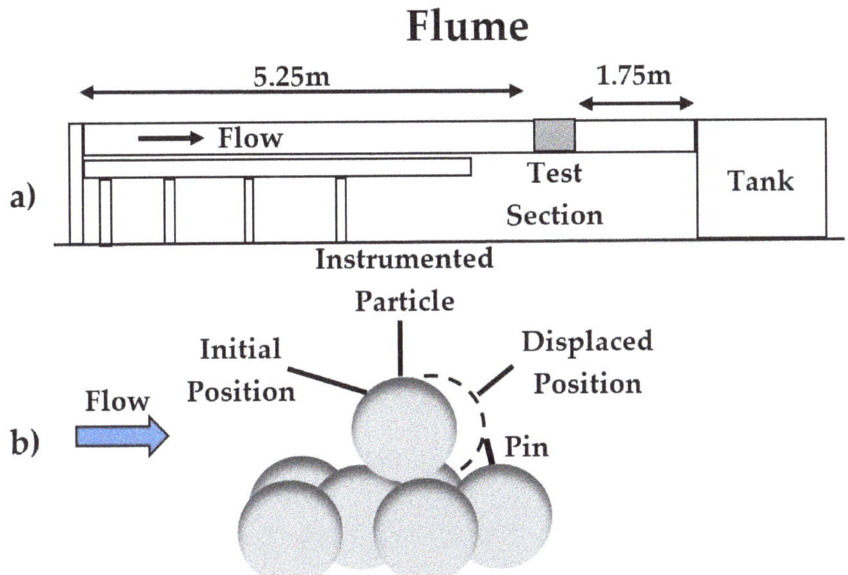

Figure 9. Schematic of the (**a**) experimental set-up showing the relative location of the test section and (**b**) test section, showing the instrumented particle resting on top of the tetrahedral bed particle arrangement (side view).

4.2. Local Microtopography

A 3D-printed bed micro-topography, consisting of four hemispheres in a rectilinear well-packed arrangement, is designed to position the spherical instrumented particle of 4 cm in diameter to study its turbulent flow induced entrainments. The particle is sitting on the local micro-topography, which is embedded within the bed surface, and is allowed to move from the upstream towards the downstream pocket due to sufficiently energetic instantaneous near bed surface flow structures. However, full entrainment is restrained due to a pin 2 mm downstream of the instrumented particle. With this method, the particle falls back to its resting location after the advection of energetic flow events pass it, thus enabling a continuous series of entrainments without disturbing the experiment by manually moving the particle back to its initial position [5,7].

4.3. The Instrumented Particle

The sensor's inertial measurement unit (IMU) is located at the centre of a 4 cm diameter spherical enclosure that maintains the electronic components waterproofed during testing, and ensures they are not subjected to excessive stresses. A range of different particle shapes and sizes of enclosures can be used depending on the objective. Having the sensor at the centre of mass, surrounded by equally distributed weights of lead, a uniform distribution of mass is obtained, so that the instrumented particle's motion is not biased and any offsets in the readings are avoided.

The casing is designed to be versatile to accommodate different amount of weights. It is important to have an adjustable density for different flow regimes, as the resistance to flow-induced entrainment is directly proportional to the particle's submerged weight. A density of 2.65 g/cm^3 is used for the testing, which is the density of the quartz earth sediment typically found in nature. The casing designs are produced using SolidWorks^{TM} and are built using a rigid opaque photopolymer material and additive manufacturing (3D-printing) equipment (Formlabs, Somerville, MA, USA).

4.4. Experimental Protocol

A high-speed video camera, operating at 120 frames per second, is placed on one side of the flume near the test section in order to record the particle's movements accurately. The waterproof camera is of minimum dimensions (2 × 2 cm) and placed slightly downstream of the test section in order to avoid any interference with the flow properties near the test section that could affect the instrumented particle's response. An ultra-bright light source is placed above the test section, and a high contrast black material is placed in the background of the camera's viewing area in order to enable accurate monitoring.

The flume is run at a flowrate that results in a continuous entrainment of the instrumented particle, which is confirmed by eye, i.e., the particle remained entrained and is stopped from moving further by the pin that is set in place. Then, the flowrate is reduced incrementally to finally have three flowrates that result in some periodic entrainments. A Nortek Vectrino1 acoustic Doppler velocimetry (ADV) is used to obtain the velocity profiles 10 cm upstream of the test section. The total time of the experiment is about 1 h, with the flow measurements being made after removing the test section to prevent any interference. Each of the three runs involves 10 min flow measurement recordings and 10 min sensor recordings. The sensor logs for about an hour at a frequency of 200 Hz, which serves as another confirmation that the instrumented particle can be deployed to record data for a sufficiently long time.

5. Results and Discussion

A proper filter, with the input of accelerometer noise and gyroscope noise values changing based on the calibration results, as discussed in Section 3.5, is used for the inertial sensor fusion. The results of sensor fusion are then used to calculate the corrected total acceleration for the three runs using Equation (1). For the three runs, the corrected total acceleration results are investigated and compared to the video recordings of the side camera. Based on a manual comparison between the sensor's data and the video recordings, a threshold value for the total acceleration is obtained by ensuring that the peaks in the total acceleration, corresponding to full entrainment events and twitching, are above and below this line, respectively. An illustration of how the corrected total acceleration results could be used to detect the range of the instrumented particle's frequencies is shown in Figure 10 below.

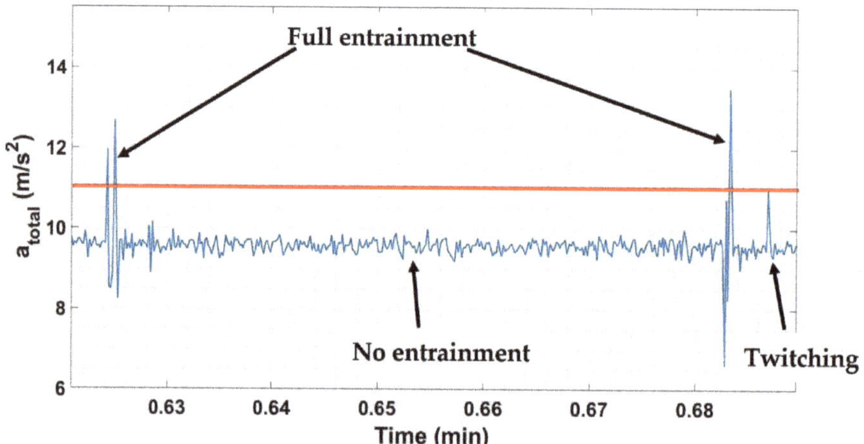

Figure 10. An illustration of how the fused sensor readings (corrected total acceleration results specifically) could be used to detect the range of the instrumented particle's frequencies. The thick red line represents a threshold-total acceleration value.

In Figure 10, a total acceleration value of almost 9.81 m/s^2 indicates no entrainment of the instrumented particle or that no energetic event acts on the instrumented particle by turbulent flowing water that is strong enough to overcome its resistance. In other words, the particle is at its resting position and the accelerometer records the gravitational acceleration in the z-direction and records zero in the x- and y-directions. At 0.688 min, the total acceleration results show a relatively small spike, indicating that there is a small energetic event(s) that could not result in the full entrainment of the instrumented particle, but rather in twitching it only. On the other hand, the total acceleration values are greater than exceedance (or the threshold-total acceleration value) at 0.625 min and 0.683 min, which indicate full entrainment events of the instrumented particle. At each instance of both events, there is an energetic event(s) that is strong enough to dislodge the instrumented particle from its resting position and move it against the restraining pin. The procedure, illustrated in Figure 10, is applied to the total acceleration results of the three runs by eliminating the values that are below a threshold-total acceleration value of 11 m/s^2 from the analysis. In other words, only full entrainment events are left to derive the frequency of entrainment of the instrumented particle for the three runs of the experiment, as shown in Figure 11 below.

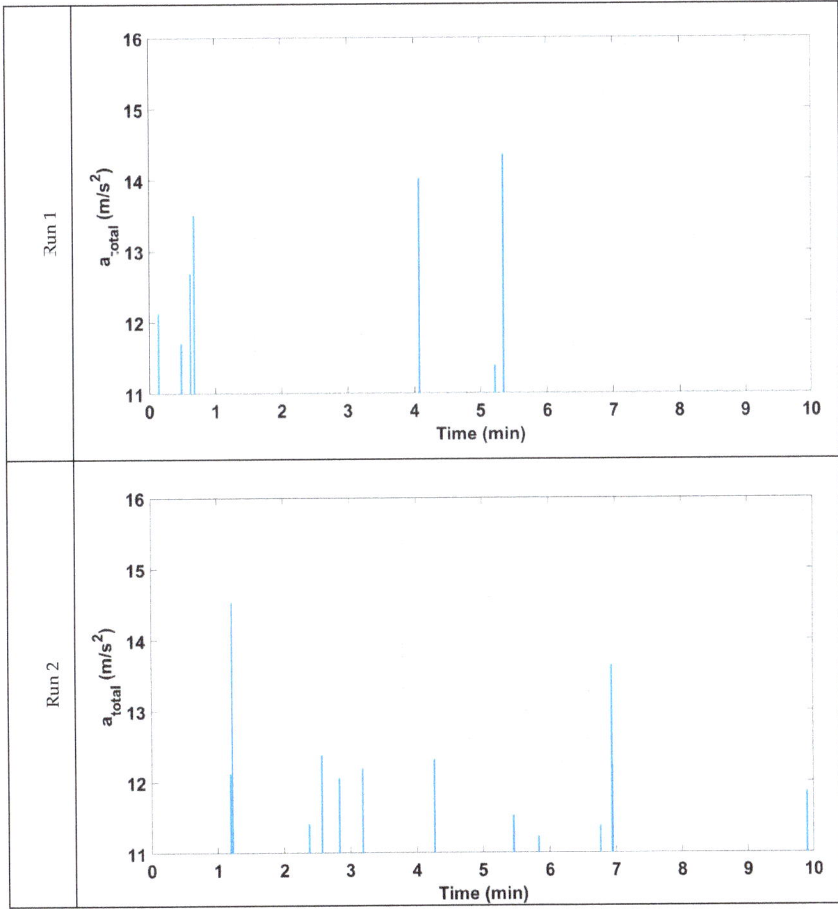

Figure 11. *Cont.*

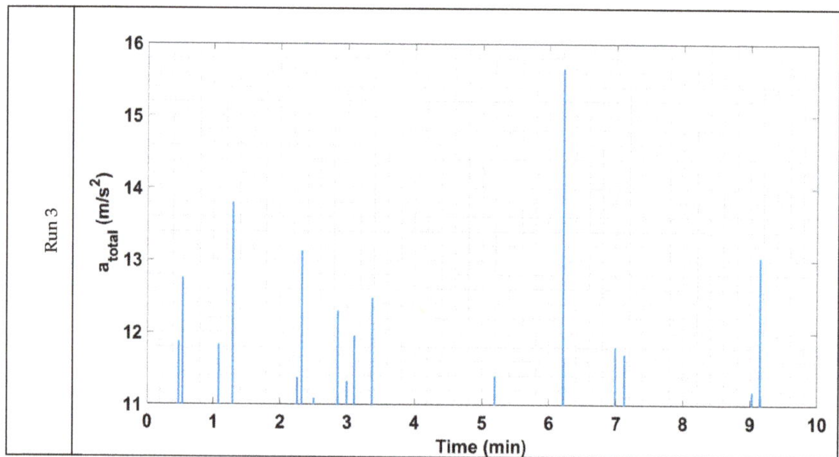

Figure 11. Above threshold-corrected total acceleration results for the three runs of the experiment.

Using the results shown in Figure 11, the frequency of entrainment, f_E, of the instrumented particle for each run is calculated by dividing the total number of full entrainments by the total run time in seconds. These calculations were performed, and the results are shown in Table 5. The results are consistent with the experimental findings of Valyrakis et al. [6] for the frequency of entrainment of a coarse particle for near-threshold conditions. As for the flow velocity measurements, they are used to estimate the shear or friction velocity (u_*) as shown in Table 5 using the Townsend method [29], which is based on the law of wall. The shear velocity, water properties, and particle diameter are used to estimate boundary, or particle, Reynolds number (Re_*) for each run as shown in Table 5. The shear velocity and water density are then used to calculate the bed shear stress (τ_o) exerted by flowing water, which is used in addition to the particle's submerged weight (W_s) to estimate Shields number or Shields parameter [30]. Shields parameter (τ_*) is a dimensionless number that is basically the ratio of fluid force exerted on the particle to the apparent or submerged weight of the particle. It is the most used criterion in the literature to determine whether there is initiation of motion of a sediment particle. Based on the results shown in Table 5, Shields parameter values for the three runs of the experiment are around the regime of incipient flow conditions predicted by a plethora of experiment studies (see [31,32], amongst others). Additionally, Nikuradse's equivalent sand roughness, which is an indicative parameter of the friction matched against an equivalent sand-grain roughness of height k_s, is estimated for the three runs of the experiment and results are presented in Table 5.

Table 5. Description and results of the three runs of the experiment.

Run Name	$Re_*=\frac{\rho u_* D}{\mu}$	u_* (mm/s)	k_s (mm)	$\tau_o=\rho u_*^2$ (Pa)	$\tau_*=\frac{\tau_o}{(\rho_s-\rho_w)gD}$	f_E (Hz)
Run1	1810.3	0.040	13.65	1.63	0.00252	0.017
Run2	1943.5	0.043	18.47	1.88	0.00290	0.028
Run3	2024.2	0.045	20.73	2.03	0.00313	0.042

It is clear from the results shown in Figure 11 and Table 5 that the frequency of entrainment of the instrumented particle, f_E, increases with increasing the flowrate (or boundary Reynold's number). Additionally, it is clear from the results presented in Table 5 that frequency of entrainment of the instrumented particle, f_E, increases with increasing mean Shields stresses. Based on Figures 10 and 11 and Table 5, the logged readings by the sensors could be fused to detect different types of motion of the particle accurately without the need for visual observations. Specifically, full entrainment, twitching, and no entrainment could be detected using the total acceleration results after defining a threshold value. Additionally, the corrected total acceleration results could be used to estimate the entrainment frequency of the instrumented particle. Such a metric could be linked to the probability of entrainment of individual sediment particles. Therefore, the instrumented particle provides a direct and cost-effective tool for monitoring the initiation of transport processes of coarse sediment particles in turbulent environments. This is of practical relevance as visual observation methods are often restricted in turbulent environments (and commonly high turbidity environments).

There are a number of tools developed to monitor the actual occurrence of scour around hydraulic infrastructure [33–35]. However, many geomorphological and hydrological hazards, such as scour around hydraulic infrastructure or riverbed and riverbank destabilisation, typically develop very fast relative to our capacity to take action once they are initiated or detected. Therefore, assessing the probability of the occurrence of the start of scour processes, before the conditions for any further critical catastrophes or infrastructure failure start setting in, is of interest.

While tools developed in the literature are designed to assess scour in real time, the tool offered herein is designed following a different approach: monitoring the risk of surface particle removal. This inherently links to well-established theories in hydraulic engineering around the probability of entrainment of individual sediment particles (see [6], amongst others). According to such probabilistic approaches, monitoring the frequency of entrainments of the most-exposed particle resting on the bed surface in the vicinity of the hydraulic infrastructure of interest should suffice to detect the risk of the onset of scour occurring before it actually takes place. In other words, monitoring the probability of particle destabilisation (or probability of entrainment) from the exceedances of acceleration, or instances where acceleration is greater than exceedance. Thus, this research focuses on the development, calibration and demonstration of the utility of the instrumented particle for assessing the probability of the onset of scour. The results from the flume experiments presented herein showcase that, by placing the instrumented particle at an appropriate location at a distance scaling with the dimension of the infrastructure (e.g., bridge pier diameter) downstream of it, offers a tool to assess the risk of scour initiation.

Additionally, performance indicators relevant to bed stability or the risk for scour can be derived from the demonstrated sensor fusion of the recordings of the inertial sensor, using, for example, metrics such as the rate of entrainment or the frequency of entrainment for the instrumented particle. This can be derived by simply estimating the number of full entrainments over the total logging period and serves as a probabilistic estimation of the risk of the destabilisation of bed material or the potential of scour development around built infrastructure. When combining such results with machine learning and algorithms, hybrid advanced data-driven predictive models can be developed in the future to assess water infrastructure risk and determine the potential for bed surface destabilisation [5,6]. Therefore, the instrumented particle has the potential to become a vital addition in the arsenal of sensors used for structural health monitoring, providing an early (well before any critical failures may appear), robust, rapid and cost-effective assessment of risks to the structural integrity of the infrastructure near water.

Furthermore, the results of the inertial sensor fusion of the logged readings could be used for estimating the near bed surface instantaneous hydrodynamic forces. Thus, it could help researchers to obtain a better understanding of the fundamental transport processes relevant to a plethora of fields, ranging from ecohydraulics to geomorphology. Additionally, experimentation with instrumented particles, as shown herein, can produce a large range of datasets, which can be used to benefit the numerical modelling of solids transport processes via validating Lagrangian particle transport models, including discrete element modelling (DEM), and can also be coupled with computational fluid dynamics (CFD-DEM) studies.

6. Conclusions

Electronic system integration at this micro-scale and with this complexity is a difficult task, however, it is greatly rewarding, as the developed instrumented particle is quite versatile and has a considerable and promising potential for future applications in the field of ecohydraulics and geosciences, as well as for all earth surface hazard sciences pursuing lab, field or numerical modelling techniques that can benefit from validating their Lagrangian particle transport models.

The newly produced instrumented particle design features a significantly reduced size (40 mm compared to past versions of instrumented particles), proving that it is entirely possible to create a miniaturised user-friendly device (instrumented particle) that allows for monitoring bed surface destabilisation or local scour in a non-intrusive and direct manner.

The calibration process for the sensors is performed with quantifying the uncertainties in the process and the sensor's readings and the results of the calibration are used to determine the input for the inertial sensor fusion filter. The sensor is also tested in a recirculating open channel with a controlled water flow, flume and the results of the fusion of the testing experiments are shown and discussed. Such a low-cost device can have a wide range of applications in water engineering and infrastructure monitoring, especially in harsh and remote environments, allowing for a better capacity for the extraction of performance indicators (such as the intensity or frequency of particle motion). The rate of entrainment as a performance indicator can, if combined with machine learning and artificial intelligence algorithms, be used to provide predictive models that can be used to statistically determine the risk of destabilisation of the bed surface.

Future work includes considering more versatile ways for transferring data using frameworks such as the Internet of Things (IoT), where data can be locally transmitted to the base where the instrumented particle rests, then out of the water body via waterproof cables and via satellite (3GPRS) and delivered to the centre of monitoring operations. Additionally, algorithms to provide real-time analysis of data, which could be of a great value, especially for assessing the risk of scour in remote and harsh environments, are to be considered. Finally, even further miniaturisation to widen the applications of the instrumented particle is currently under consideration.

Supplementary Materials: The following are available online at https://susy.mdpi.com/user/submission/video/d6a75069d08308f481caa1fccb6a803e, Video S1: Video Abstract.

Author Contributions: Conceptualization, M.V.; Data curation, K.A.-O. and Y.X.; Formal analysis, K.A.-O.; Funding acquisition, M.V.; Investigation, K.A.-O., Y.X. and M.V.; Methodology, K.A.-O. and M.V.; Project administration, M.V.; Software, K.A.-O.; Supervision, M.V.; Validation, K.A.-O.; Visualization, K.A.-O. and M.V.; Writing—original draft, K.A.-O. and M.V.; Writing—review and editing, K.A.-O., Y.X. and M.V. All authors have read and agreed to the published version of the manuscript.

Funding: This research was supported by the Royal Society (Research Grant RG2015 R1 68793/1) and the Royal Society of Edinburgh (Crucible Award).

Conflicts of Interest: The authors declare no conflict of interest.

References

1. Valyrakis, M.; Alexakis, A.; Pavlovskis, E. "Smart pebble" designs for sediment transport monitoring. In Proceedings of the 17th EGU General Assembly (EGU2015), Vienna, Austria, 12–17 April 2015.
2. Valyrakis, M.; Alexakis, A. Development of a "smart-pebble" for tracking sediment transport. In Proceedings of the International Conference on Fluvial Hydraulics (River Flow 2016), St. Louis, MO, USA, 12–15 July 2016.
3. Valyrakis, M.; Farhadi, H. Investigating coarse sediment particles transport using PTV and "smart-pebbles" instrumented with inertial sensors. In Proceedings of the 19th EGU General Assembly (EGU2017), Vienna, Austria, 23–28 April 2017.
4. Izadinia, E.; Heidarpour, M.; Schleiss, A. Investigation of turbulence flow and sediment entrainment around a bridge pier. *Stoch. Environ. Res. Risk Assess.* **2012**, *27*, 1303–1314. [CrossRef]
5. Valyrakis, M.; Diplas, P.; Dancey, C.L.; Greer, K.; Celik, A.O. Role of instantaneous force magnitude and duration on particle entrainment. *J. Geophys. Res. Earth Surf.* **2010**, *115*, 1–18. [CrossRef]
6. Valyrakis, M.; Diplas, P.; Dancey, C.L. Entrainment of coarse grains in turbulent flows: An extreme value theory approach. *Water Resour. Res.* **2011**, *47*, 1–17. [CrossRef]
7. Valyrakis, M.; Diplas, P.; Dancey, C.L. Entrainment of coarse particles in turbulent flows: An energy approach. *J. Geophys. Res. Earth Surf.* **2013**, *118*, 42–53. [CrossRef]

8. Liu, D.; Valyrakis, M.; Williams, R. Flow Hydrodynamics across Open Channel Flows with Riparian Zones: Implications for Riverbank Stability. *Water* **2017**, *99*, 720. [CrossRef]
9. Diplas, P.; Celik, A.O.; Valyrakis, M.; Dancey, C.L. Some Thoughts on Measurements of Marginal Bedload Transport Rates Based on Experience from Laboratory Flume Experiments. In *U.S. Geological Survey Scientific Investigations Report 2010–5091*; United States Geological Survey: Reston, VA, USA, 2010; pp. 130–142.
10. Koursari, E.; Wallace, S.; Valyrakis, M.; Michalis, P. Remote Monitoring of Infrastructure at Risk due to Hydrologic Hazards and Scour. In Proceedings of the 21st EGU General Assembly (EGU2019), Vienna, Austria, 7–12 April 2019.
11. Papanicolaou, T.; Knapp, D. A particle tracking technique for bedload motion. In Proceedings of the 7th International Conference on HydroScience and Engineering (ICHE2006), Philadelphia, PA, USA, 10–13 September 2006.
12. Akeila, E.; Salcic, Z.; Kularatna, N.; Melville, B.; Dwivedi, A. Testing and calibration of smart pebble for river bed sediment transport monitoring. In Proceedings of the IEEE Sensors, Atlanta Georgia, GA, USA, 28–31 October 2007.
13. Akeila, E.; Salcic, Z. Smart Pebble for Monitoring Riverbed Sediment Transport. *IEEE Sens. J.* **2010**, *10*, 1705–1717. [CrossRef]
14. Kularatnal, N.; Kularatna-Abeywardana, D. Use of motion sensors for autonomous monitoring of hydraulic environments. In Proceedings of IEEE Sensors, Lecce, Italy, 26–29 October 2008.
15. Gronz, O.; Hiller, P.; Wirtz, S.; Becker, K.; Iserloh, T.; Seeger, M.; Brings, C.; Aberle, J.; Casper, M.; Ries, J. Smartstones: A small 9-axis sensor implanted in stones to track their movements. *Catena* **2016**, *142*, 245–251. [CrossRef]
16. Frank, D.; Foster, D.; Sou, I.; Calantoni, J.; Chou, P. Lagrangian measurements of incipient motion in oscillatory flows. *J. Geophys. Res. Oceans* **2014**, *120*, 244–256. [CrossRef]
17. Zhang, Y.; Ghannam, R.; Valyrakis, M.; Heidari, H. Design and Implementation of a 3D Printed Sensory Ball for Wireless Water Flow Monitoring. In Proceedings of the 4th International Conference on UK-China Emerging Technologies, Glasgow, UK, 21–22 August 2019; pp. 1–4.
18. Maniatis, G.; Hoey, T.; Hassan, M.; Sventek, J.; Hodge, R.; Drysdale, T.; Valyrakis, M. Calculating the Explicit Probability of Entrainment Based on Inertial Acceleration Measurements, J. *Hydraul. Eng.* **2017**, *143*, 1–43. [CrossRef]
19. Noss, C.; Koca, K.; Zinke, P.; Henry, P.H.; Navaratnam, C.U.; Aberle, J.; Lorke, A. A Lagrangian drifter for surveys of water surface roughness in streams. *J. Hydraul. Res.* **2019**, 1–18. [CrossRef]
20. Houston, C.; Muir, D.; Valyrakis, M. Infrastructure scour risk assessment using instrumented particles. In Proceedings of the UK/China Emerging Technologies (UCET), Glasgow, UK, 21–22 August 2019.
21. MPU-9250 Product Specification Revision 1.1. Available online: https://www.invensense.com/wp-content/uploads/2015/02/PS-MPU-9250A-01-v1.1.pdf (accessed on 21 November 2019).
22. CoinPower CP 1654 A3 Datasheet. Available online: https://products.varta-microbattery.com/applications/mb_data/documents/data_sheets/DS63165_3.pdf (accessed on 30 January 2018).
23. DTP301120 Product Specification. Available online: http://cdn.sparkfun.com/datasheets/Prototyping/SPE-00-301120-40mah-en-1.0ver.pdf (accessed on 28 January 2018).
24. Rechargable Lithium-ion Cell. Available online: https://products.varta-microbattery.com/en/products/batteries-cells-configurations/technology/rechargeable/lithium-button-cells/all/technology-description.html (accessed on 29 January 2018).
25. Pendrill, A.; Rohlén, J. Acceleration and rotation in a pendulum ride, measured using an iPhone 4. *Phys. Educ.* **2011**, *46*, 676–681. [CrossRef]
26. Cechowicz, R. Bias Drift Estimation for MEMS Gyroscope Used in Inertial Navigation. *Acta Mech. Autom.* **2017**, *11*, 104–110. [CrossRef]
27. Xing, H.; Hou, B.; Lin, Z.; Guo, M. Modeling and Compensation of Random Drift of MEMS Gyroscopes Based on Least Squares Support Vector Machine Optimized by Chaotic Particle Swarm Optimization. *Sensors* **2017**, *17*, 2335. [CrossRef] [PubMed]
28. Kalman, R.E. A New Approach to Linear Filtering and Prediction Problems. *J. Basic Eng.* **1960**, *82*, 35–45. [CrossRef]
29. Towsend, A.A. *The Structure of Turbulent Shear Flow*, 2nd ed.; The Press Syndicate of the University of Cambridge: Cambridge, UK, 1980.

30. Shields, A. *Application of Similarity Principles and Turbulence Research to Bed-Load Movement*; California Institute of Technology: Pasadena, CA, USA, 1936.
31. Buffington, J.; Montgomery, D. A systematic analysis of eight decades of incipient motion studies, with special reference to gravel-bedded rivers. *Water Resour. Res.* **1997**, *33*, 1993–2029. [CrossRef]
32. Pähtz, T.; Clark, A. H.; Valyrakis, M.; Durán, O. The Physics of Sediment Transport Initiation, Cessation, and Entrainment Across Aeolian and Fluvial Environments. *Rev. Geophys.* **2020**, *58*, 1–58. [CrossRef]
33. Michalis, P.; Tarantino, A.; Tachtatzis, C.; Judd, D.M. Wireless Monitoring of Scour and Re-deposited Sediment Evolution at Bridge Foundations based on Soil Electromagnetic Properties. *Smart Mater. Struct.* **2015**, *24*, 1–15. [CrossRef]
34. Chen, G.; Schafer, B.P.; Lin, Z.; Huang, Y.; Suaznabar, O.; Shen, J.; Kerenyi, K. Maximum scour depth based on magnetic field change in smart rocks for foundation stability evaluation of bridges. *Struct. Health Monit.* **2015**, *14*, 86–99. [CrossRef]
35. Tang, F.; Chen, Y.; Guo, C.; Fan, L.; Chen, G.; Tang, Y. Field Application of Magnet-based Smart Rock for Bridge Scour Monitoring. *J. Bridg. Eng.* **2019**, *24*. [CrossRef]

Journal of
Sensor and Actuator Networks

Article

GP-ARX-Based Structural Damage Detection and Localization under Varying Environmental Conditions

Konstantinos Tatsis, Vasilis Dertimanis *, Yaowen Ou and Eleni Chatzi

Chair of Structural Mechanics & Monitoring, Department of Civil, Environmental and Geomatic Engineering, ETH Zürich, Stefano-Franscini-Platz 5, 8093 Zürich, Switzerland; tatsis@ibk.baug.ethz.ch (K.T.); ou@ibk.baug.ethz.ch (Y.O.); chatzi@ibk.baug.ethz.ch (E.C.)

* Correspondence: v.derti@ibk.baug.ethz.ch

Received: 2 July 2020; Accepted: 4 September 2020; Published: 8 September 2020

Abstract: The representation of structural dynamics in the absence of physics-based models, is often accomplished through the identification of parametric models, such as the autoregressive with exogenous inputs, e.g., ARX models. When the structure is amenable to environmental variations, parameter-varying extensions of the original ARX model can be implemented, allowing for tracking of the operational variability. Yet, the latter occurs in sufficiently longer time-scales (days, weeks, months), as compared to system dynamics. For inferring a "global", long time-scale varying ARX model, data from a full operational cycle has to typically become available. In addition, when the sensor network comprises multiple nodes, the identification of long time-scale varying, vector ARX models grow in complexity. We address these issues by proposing a distributed framework for structural identification, damage detection and localization. Its main features are: (i) the individual estimation of local, single-input-single-output ARX models at every operational point; (ii) the long time-scale representation of each individual ARX coefficient via a Gaussian process regression, which captures dependency on varying Environmental and Operational Conditions (EOCs); (iii) the establishment of a distributed residual generation algorithm for damage detection, which produces time-series of well-defined stationary statistics, with detected discrepancies used for damage diagnosis; and, (iv) exploitation of ARX-inferred mode shape curvatures, obtained via ARX-inferred global state-space models, of the healthy and damaged states, for damage localization. The method is assessed via application on two numerical case studies of different complexity, with the results confirming its efficacy for diagnostics under varying EOCs.

Keywords: structural health monitoring; varying environmental and operational conditions; damage detection and localization; Gaussian process regression; autoregressive with exogenous inputs; distributed sensor network; mode shape curvatures

1. Introduction

In the field of condition monitoring, model-based fault diagnosis (FD) [1–3] has been gaining ground, as a robust means for condition assessment. In contrast to other schemes, which are heavily dependent on dense instrumentation systems, the existence of a model in analytical and/or numerical form introduces inter-dependencies between the monitored quantities and, in this way, guarantees a form of redundancy [4]. When the employed model draws from first principles, e.g., a finite element representation, then this offers the added benefit of response estimation in unmeasured locations. The latter is also referred to as "virtual sensing" in recent literature [5,6]. For instance, when structural, e.g., finite element, models are coupled with observers, such as those that are based on Kalman filtering, the notion of virtual sensing can be exploited in strain estimation for fatigue assessment [7–11].

However, it is often non-trivial—or even practically infeasible—to assemble a precise system model that relies on first principles. Detailed representations using finite/discrete element models or multibody representations tend to suffer from the curse of uncertainty, tied to the required definition of modeling parameters, and will typically come with considerable computational toll. This is particularly true in the modeling of wind turbines components, where, due to the complexity of the involved geometry and materials, the establishment of a structural model forms an intricate task [12]. For such components in particular, FD plays an important role, as it could serve for damage prediction and preventive maintenance. For achieving diagnosis at reduced computation, it is, in this case, desirable to avoid the utilization of physics-based models. This work exploits availability of a parametric model structure, which is purely derived from availability of data, in the form of input/output, i.e., load/response information.

FD studies have been conducted for a diverse suite of systems including civil [13–15], mechanical [16,17], wind energy [18–23], and aerospace structures [24–28]. An important challenge in estimation originates from the exposure of such engineered systems, even if of diverse functionality, to varying Environmental and Operational Conditions (EOCs) [29–31]. Most engineered systems, and structures in particular, are subject to continually changing environments, e.g., temperature and humidity conditions. Such benign variability effects often dominate the monitored quantities, often masking the changes induced by actual damage or deterioration [32–34]. In order to account for this variability in a way that renders detection of actual damage robust, output-only or input-output schemes may be followed. In the first case, detection relies exclusively on measurements of structural response, whilst in the second case inference of a model is sought between the system output (response) and the measured EOCs. Following an output-only scheme, Bernal [35] discusses damage detection and localization in civil engineering structures in absence of information on the environment. Figueiredo et al. [36] examine different machine learning alternatives; Harmanci et al. [37] exploit an output-only autoregressive approach; Sohn et al. [38] and Dervilis et al. [39] successfully test the use of auto-associative neural networks; while, Lämsä & Kullaa [40,41] rely on factor analysis. Numerous works in the ouput-only front rely on principle component analysis (PCA) for the incorporation of EOCs [42]. Applications in FD for structures include damage identification in the form of both detection [43–45] and localization [46,47].

In this work, we propose an input–output scheme for tackling dependence of structural properties on EOC variability. Regression often forms a main tool to this end, targeting the construction of an input/output dependence function, which can be then complemented with a probabilistic treatment of the estimation error for robust outlier detection [48,49]. Alternatively, time-series representations can be adopted as models of the underlying dynamics, while explicitly incorporating dependence on influencing factors, such as EOCs. Different formulations exist to this end. Linear Parameter Varying (LPV) models utilize a two-stage approach for capturing this dependence. Firstly, local (or frozen) models are computed, which correspond to specific EOC configurations, and secondly, the parameters of the identified models are interpolated to provide a single global model [50,51], operating across EOCs. Another option lies in adoption of Functionally Pooled (FP) time-series models, which are further described in Kopsaftopoulos et al. Kopsaftopoulos et al. [52] and Sakellariou and Fassois [53]. Here, the AR coefficients are modeled as explicit functions of the EOC vector, which could for instance pertain to airspeed and angle of attack for parameters met in aerospace applications. A functional basis is used comprising bivariate polynomials, resulting as tensor products from corresponding univariate polynomials (Chebyshev, Legendre, Jacobi, and other families). Spiridonakos et al. [54] and Spiridonakos and Chatzi [55] have introduced a Polynomial Chaos basis instead in order to span the functional subspace expressing the dependence of the AR models on the EOCs [56,57] tackling variability in structural components. A further alternative pertains to use of Random Coefficient (RC) models, where the time–dependent coefficients follow a multivariate Gaussian mixture model (GMM), allowing for significant flexibility in uncertainty representation [58,59]. To further separate effects, which operate on different temporal scales, Avendaño-Valencia et al. [51] propose a hierarchical time-series model. This adopts an LPV-AR representation for capturing influences on short-term dynamics, e.g., azimuth of the rotor for operating wind turbines, while long-term influences stemming from effects such as temperature, are tackled by means of Gaussian process regression (GPR) [60]. The framework is demonstrated on the tracking of the response of the Humber bridge under long-term influence that is induced by varying acting averaged wind speeds.

Based on the above, a two-stage algorithmic process is proposed in this work, with the aim (i) to construct a data-driven model of the structure, able to reliably reconstruct the response across the full range of operation of the system in its healthy state and (ii) reliably detect and possibly localize damage, via use of a suitably defined Residual Generation Algorithm (RGA) [1,61–64] featuring appropriately selected detection thresholds. The first step requires training across a sufficient number of structural response samples that sufficiently span the "global" system response. A GP-ARX model is then constructed, which is valid across the whole range of operating conditions of the structure. Once the model is put in place, a RGA is formulated as the difference between the measured response and the noise-free part of the GP-ARX model. It is demonstrated that the RGA follows a GP-AR model: this allows for delivering the theoretical statistical thresholds and defining the range of values within which the "healthy" state lies, as a function of the EOC variable vector. Any systematic deviation from the predefined threshold zone is then a safe alert for the presence of damage. It is shown that damage detection can be accomplished even by a single GP-ARX model, e.g., a single response measurement sensor. Localization, which inevitably depends on the available sensor network, proceeds by calculating the differences between the mode shape curvatures (MSCs) of the healthy and the damaged structure.

The paper is organized, as follows: Section 2 defines the problem and delivers the theoretical difference equation between a single input-output pair, including the dependence of the input-output delay with respect to the structural response type monitored. Section 3 describes the structural identification stage, formulates the RGA and establishes the damage detection and localization procedure. The method is assessed in Section 4, via two numerical investigations of different complexity, namely a spring-mass-damper system and a shear frame reflecting a common representation of a structural system. Finally in Section 5, the results are summarized and directions for further research are given.

2. Problem Formulation

A linear, viscously damped, operationally varying structural system with n degrees of freedom (DOFs) may be represented by a second–order vector differential equation of the form

$$\mathbf{M}(\xi)\ddot{\mathbf{x}}(t) + \mathbf{C}(\xi)\dot{\mathbf{x}}(t) + \mathbf{K}(\xi)\mathbf{x}(t) = \mathbf{u}(t) \tag{1}$$

where $\mathbf{x}(t)$ is the $[n \times 1]$ vibration displacement vector, $\mathbf{M}(\xi)$, $\mathbf{C}(\xi)$ and $\mathbf{K}(\xi)$ are the real $[n \times n]$ mass, viscous damping and stiffness matrices, respectively, and $\mathbf{u}(t)$ is the $[n \times 1]$ vector of excitations.

The operational variability of the structural matrices may be attributed to diverse environmental parameters (temperature, humidity, etc.), which are herein represented by a $[L \times 1]$ vector $\xi = [\xi_1, \xi_2, \ldots, \xi_L]^T$. It is assumed that their temporal evolution is sufficiently slow as compared to the variation of the structural dynamics (e.g., sufficiently lower with respect to the lowest structural vibration mode). Subsequently, Equation (1) can be considered as a local LTI model that describes the structure during a small time interval. The two different time-scales are henceforth referred to as the long time-scale, corresponding to the temporal evolution of ξ and notated by t_l, and the short time-scale, corresponding, as usual, to the temporal evolution of $\mathbf{x}(t)$ and notated by t.

Under this setting, within the short time-scale, the parameter vector ξ may be considered as constant, which is, as a realization of a multivariate random process Ξ with joint PDF $f_{\Xi}(\xi)$. Accordingly, Equation (1) may be written as

$$\mathbf{M}(\Xi)\ddot{\mathbf{x}}(t) + \mathbf{C}(\Xi)\dot{\mathbf{x}}(t) + \mathbf{K}(\Xi)\mathbf{x}(t) = \mathbf{u}(t), \text{ with } \Xi(\omega) \in \Omega \tag{2}$$

where Ω denotes the event space of the random vector Ξ. The state-space representation of Equation (2) reads

$$\dot{\zeta}(t) = \mathbf{A}_c(\Xi)\zeta(t) + \mathbf{B}_c(\Xi)\mathbf{u}(t) \tag{3a}$$
$$\mathbf{y}(t) = \mathbf{H}_c(\Xi)\zeta(t) + \mathbf{D}_c(\Xi)\mathbf{u}(t) \tag{3b}$$

in which the state and input matrices have the standard form

$$\mathbf{A}_c(\Xi) = \begin{bmatrix} \mathbf{O}_n & \mathbf{I}_n \\ -\mathbf{M}^{-1}(\Xi)\mathbf{K}(\Xi) & -\mathbf{M}^{-1}(\Xi)\mathbf{C}(\Xi) \end{bmatrix}, \quad \mathbf{B}_c(\Xi) = \begin{bmatrix} \mathbf{0} \\ \mathbf{M}^{-1}(\Xi) \end{bmatrix} \tag{4}$$

and the output and feedforward matrices, $\mathbf{H}_c(\Xi)$ and $\mathbf{D}_c(\Xi)$, respectively, depend on the type of the monitored structural response (vibration displacement, velocity, acceleration, strain, etc.) $y(t)$, of size $[s \times 1]$.

A representation of Equation (3) in discrete-time proceeds by adopting an appropriate sampling period T_s, for which the intersample behaviour of the input can be regarded as constant (e.g., zero-order hold assumption). Subsequently

$$\zeta[k+1] = \mathbf{A}(\Xi)\zeta[k] + \mathbf{B}(\Xi)\mathbf{u}[k] \tag{5a}$$
$$\mathbf{y}[k] = \mathbf{H}(\Xi)\zeta[k] + \mathbf{D}(\Xi)\mathbf{u}[k] \tag{5b}$$

for

$$\mathbf{A}(\Xi) = e^{\mathbf{A}_c(\Xi)T_s}, \ \mathbf{B}(\Xi) = [\mathbf{A}(\Xi) - \mathbf{I}_n]\mathbf{A}_c^{-1}(\Xi)\mathbf{B}_c(\Xi), \ \mathbf{H}(\Xi) = \mathbf{H}_c(\Xi), \ \mathbf{D}(\Xi) = \mathbf{D}_c(\Xi) \tag{6}$$

Equation (5) can be transformed into a modal form as

$$\boldsymbol{\eta}[k+1] = \boldsymbol{\Lambda}(\Xi)\boldsymbol{\eta}[k] + \boldsymbol{\Pi}^T(\Xi)\mathbf{u}[k] \tag{7a}$$
$$\mathbf{y}[k] = \boldsymbol{\Psi}(\Xi)\boldsymbol{\eta}[k] + \mathbf{D}(\Xi)\mathbf{u}[k] \tag{7b}$$

with $\boldsymbol{\Lambda}(\Xi)$ corresponding to the diagonal matrix of the eigenvalues of $\mathbf{A}(\Xi)$, $\boldsymbol{\Pi}^T(\Xi)$ to the matrix of modal participation vectors and $\boldsymbol{\Psi}^T(\Xi)$ to the matrix of mode shapes. It is reminded that, for an underdamped structure, the diagonal entries of $\boldsymbol{\Lambda}(\Xi)$ arrive in complex conjugate pairs, and so do the row vectors of $\boldsymbol{\Pi}^T(\Xi)$ and the column vectors of $\boldsymbol{\Psi}^T(\Xi)$.

For convenience, we will henceforth assume that the structure is excited by a single input, e.g., $\mathbf{u}[k] \to u[k]$. Subsequently, from Equation (7), the dynamics of a single input-output pair are described by

$$\boldsymbol{\eta}[k+1] = \boldsymbol{\Lambda}(\Xi)\boldsymbol{\eta}[k] + \boldsymbol{\Pi}^T(\Xi)u[k] \tag{8a}$$
$$y_\ell[k] = \boldsymbol{\psi}_\ell(\Xi)\boldsymbol{\eta}[k] + d_\ell(\Xi)u[k] \tag{8b}$$

where $\boldsymbol{\psi}_\ell(\Xi)$ and d_ℓ are the ℓ-th row and element of $\boldsymbol{\Psi}(\Xi)$ and $\mathbf{D}(\Xi)$, respectively. The application of the $\mathcal{Z}$-transform to Equation (8) yields

$$\boldsymbol{\eta}(z) = \left[z\mathbf{I}_{2n} - \boldsymbol{\Lambda}(\Xi)\right]^{-1}\boldsymbol{\Pi}^T(\Xi)u(z) \tag{9a}$$
$$y(z) = \boldsymbol{\psi}_\ell(\Xi)\boldsymbol{\eta}(z) + d_\ell(\Xi)u(z) \tag{9b}$$

and the transfer function between $y(z)$ and $u(z)$ is

$$G_\ell(z) = \boldsymbol{\psi}_\ell(\Xi)\left[z\mathbf{I}_{2n} - \boldsymbol{\Lambda}(\Xi)\right]^{-1}\boldsymbol{\Pi}^T(\Xi) + d_\ell(\Xi) \tag{10}$$

or, as a modal decomposition

$$G_\ell(z) = \sum_{i=1}^{2n} \frac{\boldsymbol{\psi}_{\ell,i}(\Xi)\boldsymbol{\pi}_i^T(\Xi)}{z - \lambda_i(\Xi)} + d_\ell(\Xi) = \sum_{i=1}^{2n} \frac{R_{\ell,i}(\Xi)z^{-1}}{1 - \lambda_i(\Xi)z^{-1}} + d_\ell(\Xi) \tag{11}$$

with $R_{\ell,i}(\Xi)$ obviously defined. In deriving a rational expression for $G(z)$, notice that the far right hand side of Equation (11) can be written as

$$G_\ell(z) = \sum_{i=1}^{2n} \frac{R_{\ell,i}(\Xi)z^{-1}\overline{A}_i(z)}{A(z)} + d_\ell(\Xi) \tag{12}$$

for

$$\begin{aligned} A(z) &= (1-\lambda_1(\Xi)z^{-1})(1-\lambda_1(\Xi)z^{-1})\dots(1-\lambda_{2n}(\Xi)z^{-1}) \\ &= 1 + a_1(\Xi)z^{-1} + a_2(\Xi)z^{-2} + \dots + a_{2n}(\Xi)z^{-2n} \end{aligned} \tag{13}$$

The polynomial $\overline{A}_i(z)$ in the numerator of Equation (12) corresponds to $A(z)$ with the i-th eigenvalue removed and it reaches up to order $-2n+1$. By combining further (i) the presence of the unit delay, z^{-1}; and, (ii) the term $d_\ell(\Xi)$, a qualitative rational expression for $G(z)$ can be derived

$$G_\ell(z) = \frac{B(z)}{A(z)} = \frac{b_0(\Xi) + b_1(\Xi)z^{-1} + b_2(\Xi)z^{-2} + \dots + b_{2n}(\Xi)z^{-2n}}{1 + a_1(\Xi)z^{-1} + a_2(\Xi)z^{-2} + \dots + a_{2n}(\Xi)z^{-2n}} \tag{14}$$

with $b_0(\Xi) = d_\ell(\Xi)$. Based on Equations (10)–(13), it follows that the coefficients of $B(z)$ depend on both of the eigenvalues and the eigenvectors of $\mathbf{A}(\Xi)$. Also observe that, in the absence of a feedforward term in Equation (8), then $G(z)$ has a unit input-output delay, since $d_\ell(\Xi) = 0$. In either case, the difference equation that describes the structural dynamics of the ℓ-th response in discrete-time stems from Equation (14) as

$$\begin{aligned} y_\ell[k] + a_1(\Xi)y_\ell[k-1] + \dots + &a_{2n}(\Xi)y_\ell[k-2n] \\ &= b_0(\Xi)u[k] + b_1(\Xi)u[k-1] + \dots + b_{2n}(\Xi)u[k-2n] \end{aligned} \tag{15}$$

Under this setting, the problem treated herein is the detection and the localization of damage when

- there is no prior information about the structure,
- input-output measurements are consistently recorded over the short scale (e.g., k), and
- the complete set of Ξ is available (measurable) over the long time-scale (e.g., t_l)

The proposed strategy is of a decentralized nature, in the sense that estimation is first accomplished on a local basis, by considering all individual input-output data pairs independently, as dictated by Equation (15). The adopted framework comprises two phases, e.g., the structural identification phase and the damage detection and localization phase, which are discussed in the following.

3. The Methodological Framework

3.1. The Structural Identification Phase

For simplicity, it is henceforth assumed that the operational variability is due to a single parameter, e.g., $L = 1$ and $\Xi \to \Xi$. Equation (15) indicates that at the healthy state of the structure, an excitation-response pair can be effectively described in discrete-time by an ARX model of the form [65]

$$y[k] + \sum_{i=1}^{p} a_i(\Xi)y[k-i] = \sum_{i=0}^{p} b_i(\Xi)u[k-i] + e[k] \tag{16}$$

where $y[k]$ denotes the noise-corrupted structural vibration response of a single DOF, $u[k]$ is the excitation, $a_i(\Xi)$ and $b_i(\Xi)$ are the coefficients of the AR and exogenous polynomials, of order p, and $e[k] \sim \mathcal{N}(0, \sigma_e^2(\Xi))$ is a zero-mean Gaussian white noise realization that models process and measurement uncertainty (e.g., model inconsistencies and instrumentation noise). It is noted that, in Equation (16), $b_0 \neq 0$ only when vibration acceleration response measurements are available.

Equation (16) indicates that the ARX(p, p) model accounts for the variability of the response due to variations of Ξ through the AR and exogenous parameters, $\alpha_i(\Xi)$ and $b_i(\Xi)$, respectively, as well as through the noise variance matrix $\sigma_e^2(\Xi)$. These are herein admitting a GPR representation as

$$a_i(\Xi) = \sum_{j=1}^{p_a} \alpha_{ij} S_{a,j}(\Xi) + P_a(\Xi) + \epsilon_a \tag{17a}$$

$$b_i(\Xi) = \sum_{j=1}^{p_b} \beta_{ij} S_{b,j}(\Xi) + P_b(\Xi) + \epsilon_b \tag{17b}$$

$$\sigma_e^2(\Xi) = \sum_{j=1}^{p_\sigma} \gamma_{ij} S_{\sigma,j}(\Xi) + P_\sigma(\Xi) + \epsilon_\sigma \tag{17c}$$

In light of this parametric representation, the model of Equation (16) is termed as GP-ARX(p, p). In Equation (17), $S_{(\cdot),(\cdot)}(\Xi)$ denotes an appropriate set of basis functions, $P_{(\cdot)}(\Xi) \sim \mathcal{GP}(0, \kappa_{(\cdot)}(\Xi, \Xi'))$, where $\kappa_{(\cdot)}(\Xi, \Xi'))$ is the kernel function, being fully determined by a set of hyperparameters $\boldsymbol{\vartheta}_{(\cdot)}$, and $\epsilon_{(\cdot)} \sim \mathcal{N}(0, \sigma^2_{\epsilon_{(\cdot)}})$

Upon proper selection of the basis and kernel functions, $S_{(\cdot),(\cdot)}(\Xi)$ and $\kappa_{(\cdot)}(\Xi, \Xi'))$, respectively, the representation of the structure through GP-ARX(p, p) models is succeeded through the solution of an associated identification problem, in which all unknown model parameters are estimated, which is, the integers p, p_a, p_b, and p_σ, the GPR coefficients α_{ij}, β_{ij} and γ_{ij} in Equation (17), as well as the vectors of hyperparameters $\boldsymbol{\vartheta}_a$, $\boldsymbol{\vartheta}_b$ and $\boldsymbol{\vartheta}_\sigma$ and the variances $\sigma^2_{\epsilon_a}$, $\sigma^2_{\epsilon_b}$, and $\sigma^2_{\epsilon_\sigma}$. Yet, since the operational parameter Ξ varies slowly, this problem cannot be treated by pooling a large regression problem over a set of Q realizations of Ξ [65,66]: in applications involving large infrastructures (bridges, buildings, wind turbines, etc.), Ξ usually varies seasonally. Thus, it is rather impractical to wait until a full operational cycle becomes available before estimating a global GP-ARX model.

In contrast, the strategy chosen herein pertains to (i) the estimation of short time-scale ARX(p, p) models, whenever data become available; (ii) the storage of the estimated AR and exogenous parameters, the estimated variance and measurement(s) of Ξ; and, (iii) the GP modelling of all parameters according to Equation (17), when a full operational cycle is completed. Notice that the direct modeling of the parameters $a_i(\Xi)$, $b_i(\Xi)$, and $\sigma_e^2(\Xi)$ in Equation (17), is a process that can be easily realized in parallel and essentially corresponds to typical data-driven uncertainty quantification problem.

The short time-scale ARX modelling step is conventionally accomplished: for any given pair of input-output data at a specific long time-scale instant t_l, $u_{t_l}[k]$ and $y_{t_l}[k]$, respectively, for $k = 1, 2, \ldots, N$, the original ARX model is transformed into its linear regression form

$$y_{t_l}[k] = \boldsymbol{\phi}_{t_l}[k]^T \boldsymbol{\theta}_{t_l} + e_{t_l}[k] \tag{18}$$

for

$$\boldsymbol{\phi}_{t_l}[k] = \Big[-y_{t_l}[k-1] \; \ldots \; -y_{t_l}[k-p] \mid u_{t_l}[k] \; \ldots \; u_{t_l}[k-p] \Big]^T \tag{19}$$

and

$$\boldsymbol{\theta}_{t_l} = [\alpha_{1,t_l} \; \alpha_{2,t_l} \; \ldots \; \alpha_{p,t_l} \mid b_{0,t_l} \; b_{1,t_l} \; \ldots \; b_{p,t_l}]^T \tag{20}$$

For $k = 1, \ldots, N$, Equation (18) implies

$$\mathbf{Y}_{t_l} = \mathbf{\Phi}_{t_l}\boldsymbol{\theta}_{t_l} + \mathbf{E}_{t_l} \tag{21}$$

where $\mathbf{Y}_{t_l} = \left[y_{t_l}[1] \ \ldots \ y_{t_l}[N]\right]^T$, $\mathbf{\Phi}_{t_l} = \left[\boldsymbol{\phi}_{t_l}^T[1] \ \ldots \ \boldsymbol{\phi}_{t_l}^T[N]\right]^T$ and $\mathbf{E}_{t_l} = \left[e_{t_l}[1] \ \ldots \ e_{t_l}[N]\right]^T$. Minimization of the quadratic least squares criterion provides the estimate

$$\hat{\boldsymbol{\theta}}_{t_l} = \left(\mathbf{\Phi}_{t_l}^T\mathbf{\Phi}_{t_l}\right)^{-1}\mathbf{\Phi}_{t_l}^T\mathbf{Y}_{t_l} \tag{22}$$

while the prediction errors are recovered by $\mathbf{E}_{t_l} = \mathbf{Y}_{t_l} - \mathbf{\Phi}_{t_l}\hat{\boldsymbol{\theta}}_{t_l}$ and their variance is calculated as $\sigma_{e,t_l}^2 = \mathbf{E}_{t_l}^T\mathbf{E}_{t_l}$.

This procedure is repeated over the whole range of available operational realizations Ξ_{t_l}, $t_l = 1, \ldots Q$. Under the reasonable assumption that the order of the structural system remains unaltered, the AR and exogenous polynomial orders, estimated statistically from an arbitrary single realization, can be maintained the same for every t_l. Subsequently, the GPR estimation of each individual parameter of Equation (17) is taking place while using the hybrid optimization method described in Dertimanis et al. [67].

3.2. The Damage Detection and Localization Phase

The successful completion of the previous phase provides the availability of s single-input, single-output GP-ARX$_\ell(p,p)$ models, for $\ell = 1, \ldots, s$, where s corresponds to the number of available DOFs' measurements. These models are capable of tracking the evolution of the structural response of the relative DOF during the healthy structural state, for the whole expected range of variations in Ξ. In establishing an effective damage detection scheme, a residual generation algorithm (RGA) is formulated as shown in Figure 1, for every AFS-ARX model. The RGA is a computational machine that operates in parallel to the actual structure and issues an alert when abnormal behavior appears, i.e., different to the one estimated during the structural identification phase.

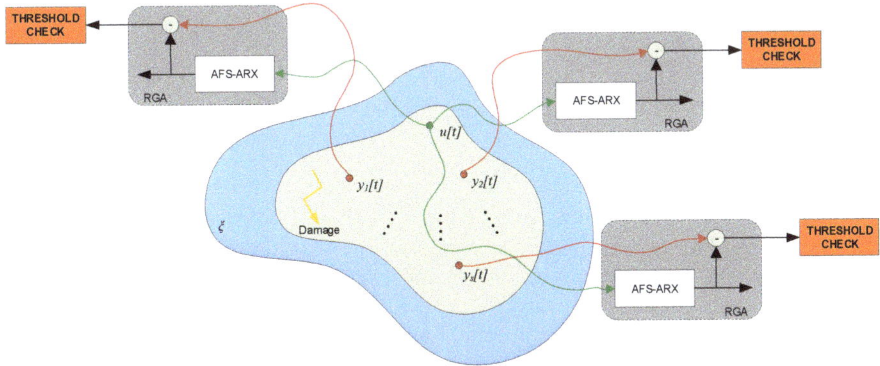

Figure 1. Layout of the residual generation algorithms for damage detection.

In order to formulate the RGA, the ℓ-th AFS-ARX model of Equation (16) is compactly written as

$$y_\ell[k] = \frac{B_\ell(q,\Xi)}{A(q,\Xi)}u[k] + \frac{1}{A(q,\Xi)}e_\ell[k] \tag{23}$$

where q is the backshift operator (i.e., $q^{-i}y[k] = y[k-i]$) and $A(q,\Xi)$, $B_p(q,\Xi)$ the AR end exogenous polynomials, respectively, of the form,

$$A(q,\Xi) = 1 + a_1(\Xi)q^{-1} + \cdots + a_p(\Xi)q^{-p} \quad (24a)$$

$$B_\ell(q,\Xi) = b_{0,\ell}(\Xi) + b_{1,\ell}(\Xi)q^{-1} + \cdots + b_{p,\ell}(\Xi)q^{-p} \quad (24b)$$

Notice that the first term of the right-hand side of Equation (23) corresponds to the estimated discrete-time transfer function between the applied excitation and the vibration response at the monitored DOF of the structure. To this end, the RGA can be simply formulated as the difference

$$r_\ell[k] = y_\ell[k] - \frac{B_\ell(q,\Xi)}{A(q,\Xi)}u[k] \quad (25)$$

retaining the units of the monitored quantity. In the absence of damage, the residual $r_\ell[k]$ is thus equal to

$$r_\ell[k] = \frac{1}{A(q,\Xi)}e_\ell[k] \quad (26)$$

implying that it follows an GP-AR(p) model of the form

$$r_\ell[k] + \sum_{i=1}^{p} a_i(\Xi)r_\ell[k-i] = e_\ell[k] \quad (27)$$

The variance of $r_\ell[k]$ can be calculated by truncating the last expression to an infinite moving average (MA) model

$$r_\ell[k] = e_\ell[k] + \sum_{i=1}^{\infty} \mathcal{H}_{i,\ell}e_\ell[k-i] \quad (28)$$

in which $\mathcal{H}_{i,\ell}$ are the coefficients of the associated Green function [68]. The mean and the variance of the time-series of Equation (28) are given by

$$\mu_{r_\ell} = E\{r_\ell[k]\} = 0 \text{ and } \sigma^2_{r_\ell}(\Xi) = E\{r^2_\ell[k]\} = \sum_{i=0}^{\infty} \mathcal{H}^2_{i,\ell}\sigma^2_{e,\ell}(\Xi) \quad (29)$$

respectively. This means that 99.7% of the damage-free RGA time-series conform to

$$-3\sigma_{r_\ell}(\Xi) \leq r_\ell[k] \leq 3\sigma_{r_\ell}(\Xi) \quad (30)$$

Equations (28)–(30) imply that the quantitative evaluation of the damage detection thresholds require knowledge of the infinite MA model coefficients $\mathcal{H}_{i,\ell}$. These are calculated in closed form via the residue expansion of the transfer function of Equation (27). The reader is referred to Box et al. [68] [Chapter 3] for further details. Here, it is simply reminded that the AR polynomial should be of minimum-phase for the whole range of operating conditions (e.g., for all values of the temperature), in order to ensure the convergence of $\mathcal{H}_{i,p}$, and that AR roots' amplitudes close to unity result in a slower the decay rate of $\mathcal{H}_{i,p}$.

Damage localization follows the detection stage and operates on data that are safely considered as outcomes of the damaged structure, e.g., after the damage time instant detected by the RGAs. Localization requires the estimation of a new set of simple ARX models, one for each available input–output data set and it is based on the differences between the mode shape curvatures between the GP-ARX models and the estimated-based-on-damaged-data ARX ones. For both sets of models, mode shape curvature estimation proceeds by first establishing the state-space realizations of the noise-free parts of the GP-ARX and ARX models [69]

$$\zeta[t+1] = \mathbf{A}\zeta[t] + \mathbf{B}u[t] \tag{31a}$$
$$\mathbf{y}[t] = \mathbf{H}\zeta[t] + \mathbf{D}u[t] \tag{31b}$$

where

$$\mathbf{A} = \begin{bmatrix} 0 & 1 & \dots & 0 \\ 0 & 0 & \dots & 0 \\ \vdots & \vdots & \ddots & 0 \\ 0 & 0 & 0 & 1 \\ -a_p & -a_{p-1} & \dots & -a_1 \end{bmatrix}, \quad \mathbf{B} = \begin{bmatrix} 0 \\ 0 \\ \vdots \\ 0 \\ 1 \end{bmatrix} \tag{32a}$$

$$\mathbf{H} = \begin{bmatrix} b_p - a_p b_0 & b_{p-1} - a_{p-1} b_0 & \dots & b_1 - a_1 b_0 \end{bmatrix}, \quad \mathbf{D} = \begin{bmatrix} b_0 \end{bmatrix} \tag{32b}$$

and then transforming Equation (31) to the modal form of Equation (8) via the eigenvalue problem of the state matrix **A**. The real part of the estimated row vector $\boldsymbol{\psi}_\ell$ (see Equation (8)) contains the values of the identified mode shapes at the ℓ-th sensor location. Gathering together all of the estimated $\boldsymbol{\psi}_\ell$ vectors, the matrix of mode shapes is constructed as

$$\boldsymbol{\Psi} = \begin{bmatrix} \boldsymbol{\psi}_1 \\ \boldsymbol{\psi}_2 \\ \vdots \\ \boldsymbol{\psi}_s \end{bmatrix} = \begin{bmatrix} \tilde{\boldsymbol{\psi}}_1 & \tilde{\boldsymbol{\psi}}_2 & \vdots & \tilde{\boldsymbol{\psi}}_p \end{bmatrix} \qquad ([s \times p]) \tag{33}$$

and its columns correspond to the shapes of each vibration mode. In establishing $\boldsymbol{\Psi}$, attention should be paid to the proper distinction of structural from "erroneous" modes, as well as to the matching of eigenvectors associated to the same mode. This is accomplished by applying dispersion analysis [70–72] and formulating the modal assurance criterion (MAC).

When $\boldsymbol{\Psi}$ is set up to contain the structural vibration modes from all of the identified ARX models, the entries of the ℓ-th column of the mode shape curvature matrix are calculated by [73]

$$[\tilde{\boldsymbol{\psi}}''_\ell]_i = \frac{[\tilde{\boldsymbol{\psi}}_\ell]_{i+1} - 2[\tilde{\boldsymbol{\psi}}_\ell]_i + [\tilde{\boldsymbol{\psi}}_\ell]_{i-1}}{h^2} \tag{34}$$

with h denoting the distance between two successive measurement nodes. Finally, damage localization is examined by the difference

$$\Delta\tilde{\boldsymbol{\psi}}''_\ell = \tilde{\boldsymbol{\psi}}''_{\ell,H} - \tilde{\boldsymbol{\psi}}''_{\ell,D} \tag{35}$$

where $\tilde{\boldsymbol{\psi}}''_{\ell,H}$ and $\tilde{\boldsymbol{\psi}}''_{\ell,D}$ correspond to the ℓ-th MSC matrix columns established by the GP-ARX models and the estimated-based-on-damaged-data ARX ones, respectively.

4. Case Studies

4.1. Damage Detection on a Spring-Mass-Damper System

A simple structural system with six DOFs, as illustrated in Figure 2, is chosen as the first case study. It is assumed that the stiffness is exponentially dependent on temperature as $K_i[t_l] = 10^{5-3K_T(T[t_l]-T_{ref})}$, where t_l corresponds to long (e.g., sufficiently lower with respect to the lowest structural vibration mode) time-scale, $K_T = 5 \times 10^{-3}$ is the thermal coefficient, $T_{ref} = 15\ ^\circ\mathrm{C}$ designates a reference temperature, and $T[t_l]$ a discrete-time stochastic process described as

$$T[t_l] = T_{ref} + T_{span} sin(2\pi f_l t_l) + w_T[t_l] \tag{36}$$

in which T_{span} = 20 °C, f_l is the long time-scale frequency and $w_T[t_l] \in \mathcal{U}(0,2)$. A stiffness proportional damping is adopted as $c_i = 2 \times 10^{-3} K_i$. It should be noted that this is a fictitious equation, which is nonetheless chosen to represent the decrease in stiffness with increasing temperature, met in most civil engineering works. Figure 3 displays a realization of $T[t_l]$ for $t_l = 1, 2, \ldots, 365$ and $f_l = 1/365$, along with its sample and kernel-based probability density function, respectively. For this realization, the induced stiffness and structural vibration modes (natural frequencies and damping ratios) are shown in Figures 4 and 5, respectively.

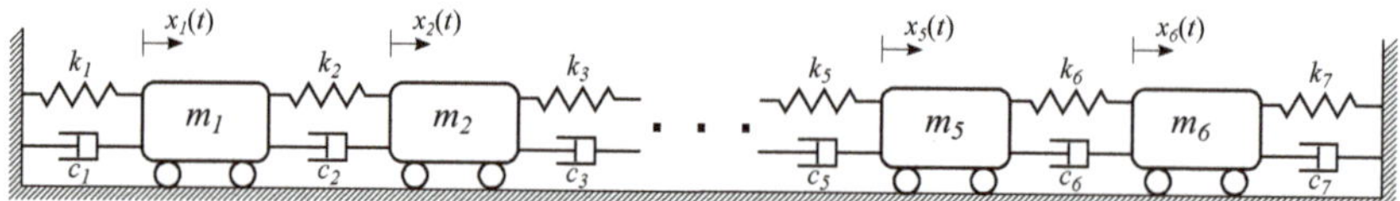

Figure 2. 6-degrees of freedom (DOF) structural system with $m_i = 10^2$ kg, $c_i = 2 \times 10^2$ Ns/m and temperature-varying stiffness K_i.

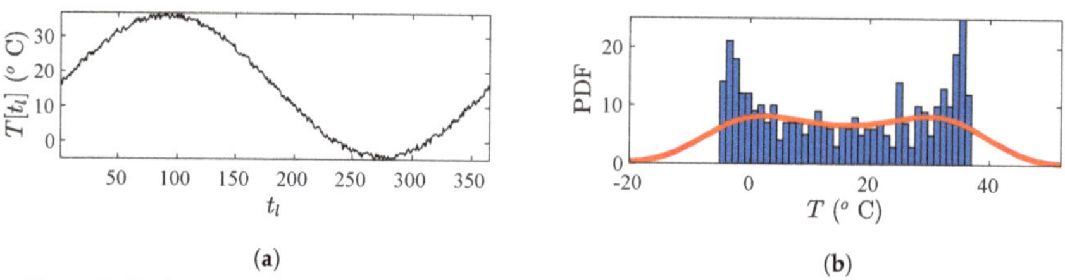

Figure 3. Spring-mass-damper: a realization of the temperature stochastic process: (**a**) Time-series. (**b**) Sample (blue bars) and associated kernel-based (continuous red line) PDF estimates.

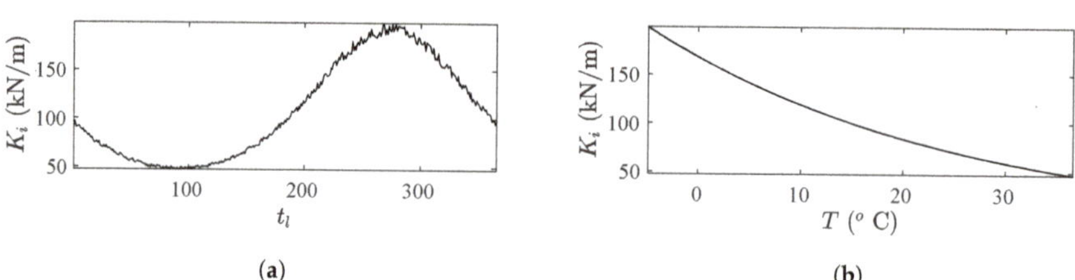

Figure 4. Spring-mass-damper: structural stiffness for the temperature realization of Figure 3. (**a**) K_i vs. t_l. (**b**) K_i vs. T.

Under a single excitation acting at DOF #3, all instances of the "healthy" structure are discretized at $T_s = 0.02$ s ($F_s = 50$ Hz) via the zero-order hold principle. Accordingly, the structural vibration acceleration responses of the discretized instances $\mathcal{S}_d[t_l]$ under independent realizations of zero-mean, Gaussian white noise excitation ($u[k] \in \mathcal{N}(0, 1.87 \times 10^4)$, and data length $N = 2^{14}$) are obtained, zero-mean subtracted and noise-corrupted at 1% noise-to-signal-ratio. For the subsequent analysis, it is assumed that only measurements from DOFs #1, #3, and #5 are available. For these, Figure 6 displays the amplitudes of the estimated frequency response functions, along with their theoretical counterparts, while Figure 7 shows the driving point FRF at reference temperature.

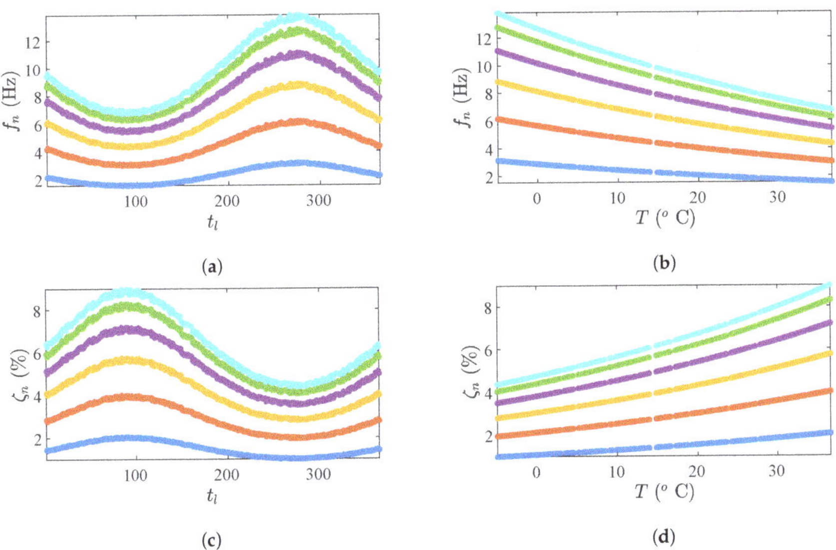

Figure 5. Spring-mass-damper: structural vibration modes for the temperature realization of Figure 3: (**a**) f_n vs. t_l. (**b**) f_n vs. T. (**c**) ζ_n vs. t_l. (**d**) ζ_n vs. T.

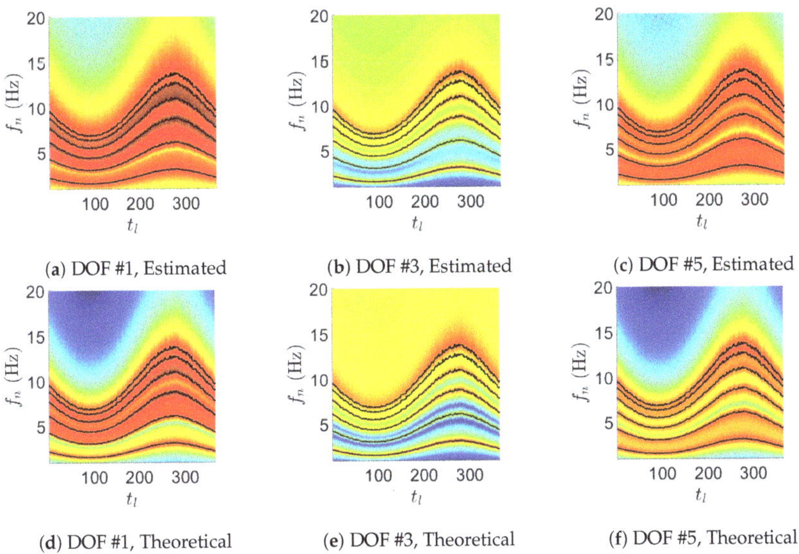

Figure 6. Spring-mass-damper: estimated (top row, Welch's method with Hanning window, $N_{FFT} = 2^{11}$, 50% overlap) and theoretical (bottom row) amplitudes of the frequency response functions at measured DOFs. Continuous black curves correspond to the long time-varying natural frequencies.

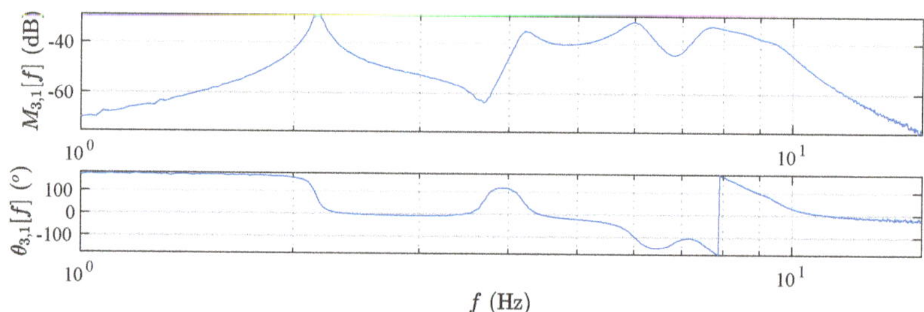

Figure 7. Spring-mass-damper: driving point FRF at reference temperature.

The establishment of GP-ARX models for describing the input–output dynamics at measured DOFs proceeds by (i) determining an appropriate order for the AR and exogenous polynomials; (ii) estimating ARX models for the whole long time-scale estimation set, consisting of the first 300 realizations (e.g., $t_l = 1, \ldots, 300$); (iii) performing GPR modelling for all the associated AR and exogenous polynomials' coefficients, as well as the variance of the prediction errors; and, (iv) assessing the resulted AFS-ARX models by applying the RGA to the validation set.

Model order selection is accomplished by considering the driving point measurement at a randomly chosen long time instant $t_r \in [1, 300]$, e.g., $\mathcal{S}_d[t_r]$. To this end, ARX(p,p) models with zero delay are estimated for $p = 10, 12, \ldots, 50$ and compared while using the Bayesian Information Criterion (BIC) and the Mean Squared Error (RSS), defined as

$$BIC = \ln\left(\sigma_e^2\right) + (2p+1)\frac{\ln\left(N_e\right)}{N_e} \tag{37a}$$

$$MSE = \frac{1}{N_e}\sum_{k=1}^{N_e} e^2[k] \tag{37b}$$

respectively. In Equation (37), N_e corresponds to the short time-scale estimation set, herein defined as the subset of $N = 1, \ldots, 16{,}384$ that lies in [1010, 15,565].

The performance of the model order selection criteria is depicted in Figure 8a,b. The BIC and the MSE both decay slowly and stabilize at high orders ($p > 30$). To elaborate further on the model order selection, Figure 8c shows the frequency stabilization diagrams (FSDs), corrected using the dispersion analysis method described in Dertimanis and Chatzi [71]. The correction is based on the energy associated to each estimated structural vibration mode and on a low threshold (herein set as greater than 1% of the mode with the highest energy content) that distinguishes structural from erroneous, or "artificial" modes [70]. The dispersion-corrected FSD indicates that mode and dispersion stabilization occurs at $p = 38$, which is the finally selected order for all three ARX models. Figure 9 illustrates the behaviour of the ARX3(38, 38) prediction errors, from where the hypothesis of Gaussian white noise process can be safely adopted.

Based on these results, ARX$1_{,t_l}(38,38)$ (unit delay), ARX$3_{,t_l}(38,38)$ (zero delay), and ARX$5_{,t_l}(38,38)$ (unit delay) models are subsequently estimated for $t_l = 1, \ldots, 300$ and their parameter vectors and prediction error variances are stored for the GPR modelling stage. As Table 1 indicates, a quite satisfying degree of consistency may characterize the parametric identification process, since the percentage fitness

$$fit = 100\left(1 - \frac{||y[t] - \hat{y}[t]||}{||y[t] - \mu_y||}\right) (\%) \tag{38}$$

to the short time estimation data set is sufficiently high.

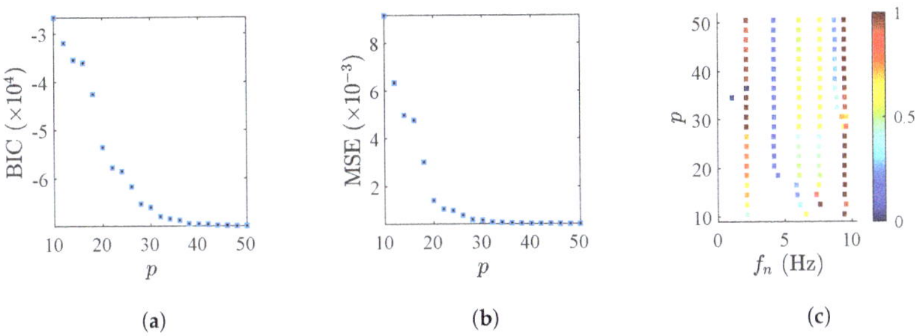

(**a**) (**b**) (**c**)

Figure 8. Spring-mass-damper: performance of the ARX model order selection criteria: (**a**) BIC. (**b**) MSE. (**c**) Dispersion-corrected frequency stabilization diagram.

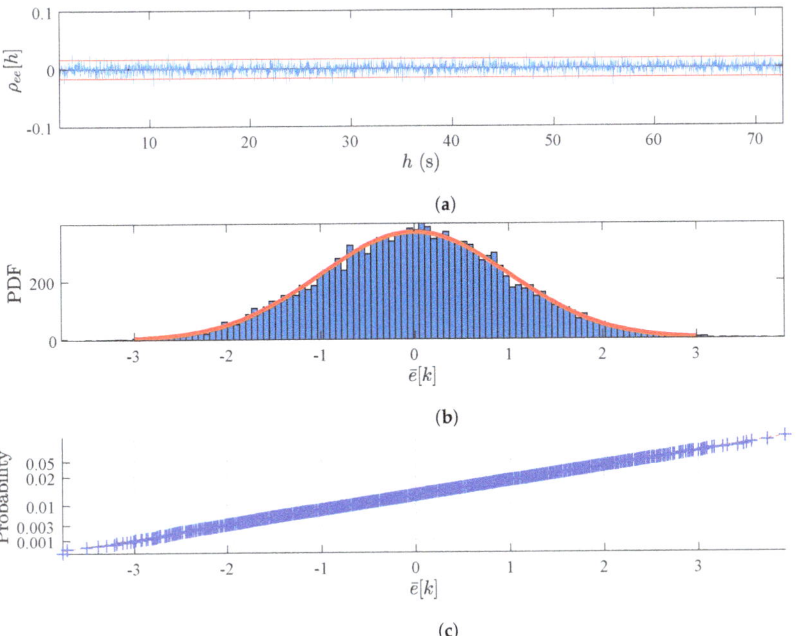

(**c**)

Figure 9. Spring-mass-damper: performance of the $ARX_3(38, 38)$ prediction errors: (**a**) Sample autocorrelation function. (**b**) Sample PDF of the normalized ($\bar{e}[k] \in \mathcal{N}(0, 1)$) prediction errors. (**c**) Normal probability plot of $\bar{e}[k]$.

Table 1. Spring-mass-damper: statistics of percentage fitness to the short time estimation data set during ARX modelling.

	$\mathbf{ARX_{1,t_l}(38, 38)}$	$\mathbf{ARX_{3,t_l}(38, 38)}$	$\mathbf{ARX_{5,t_l}(38, 38)}$
μ_{fit}	98.91%	98.93%	98.91%
σ_{fit}	4.2×10^{-2}	1.9×10^{-2}	5.0×10^{-2}

Following the structural identification stage, GPR models in the form of Equation (17) are fitted to a total number of 228 parameters. It is noted that no modelling is performed for the b_0 parameters: this is due to the fact that the AFS-$ARX_1(38, 38)$ and AFS-$ARX_5(38, 38)$ models are characterized by unit delay, while the b_0 parameter of the AFS-$ARX_3(38, 38, 0)$ model is consistently estimated at 0.01 during the structural identification stage, with negligible fluctuations around this value.

The selected basis functions and GPR orders, determined via an initial trial procedure, are listed in Table 2, while the percentage fitness results are shown in Figure 10 and Table 3. In general, sufficiently high fitting to the parameters has occurred, reaching, in many cases, more than 99%. There exist some AR and eXogenous parameters with low modelling quality, as for example the a_3 of GP-ARX3(38,38) (Figure 10a, second row) and the b_{34} and b_{36} of the same model (Figure 10b, second row); these exhibit wider scattering as compared to others, therefore rendering the fitting process more demanding. Still, an effective GPR representation for these is succeeded. This can be visualized in Figure 11, where indicative behaviour of the fitted GPR models is shown. In specific, the associated plots correspond to AR, eXogenous and variance parameter models with the lowest and highest percentage fitness.

Table 2. Spring-mass-damper system: GPR basis functions, kernels and orders.

Quantity	AR Parameters	Exogenous Parameters	Noise Variance
Basis Function Type	Discrete Fourier		Polynomial
Basis Function Formula	$S_{a,j}(\Xi) = \cos\left(\frac{2\pi j}{300}\Xi\right), j = 0,2,4,\ldots$ $S_{a,j}(\Xi) = \sin\left(\frac{2\pi j}{300}\Xi\right), j = 1,3,5,\ldots$		$S_{b,j}(\Xi) = \Xi_j, j = 0,1,2,\ldots$
Kernel Type		Matérn 3/2	
Kernel Formula		$\kappa(\Xi_i, \Xi_j) = \sigma^2\left(1 + \frac{\sqrt{3}(\Xi_i - \Xi_j)}{\rho}\right)e^{-\frac{\sqrt{3}(\Xi_i,\Xi_j)}{\rho}}$	
Order	$p_a = 3$	$p_b = 17$	$p_\sigma = 3$

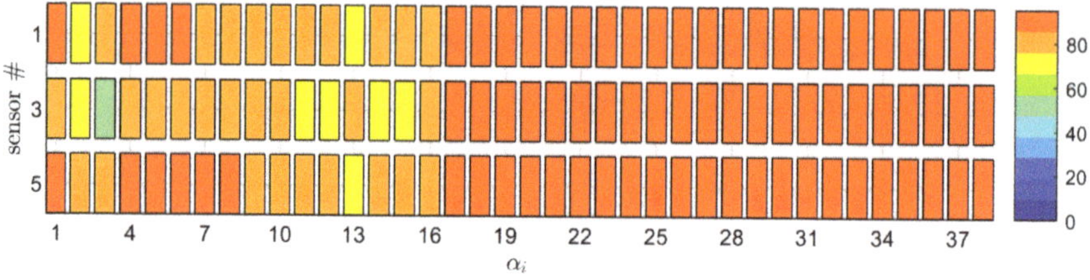

(a) AR parameters

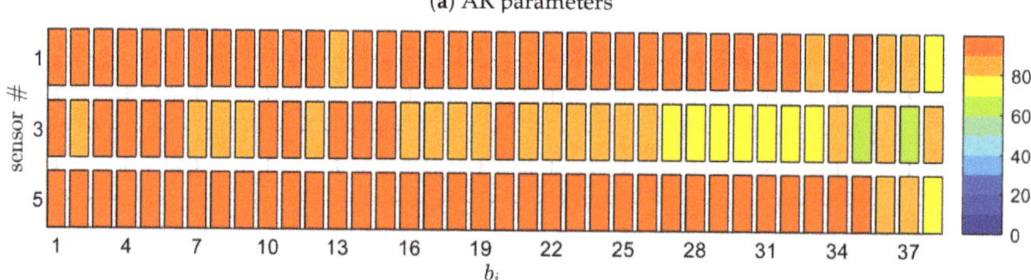

(b) eXogenous parameters

Figure 10. Spring-mass-damper: GPR modelling results for the AR and the eXogenous parameters of the associated ARX models.

Table 3. Spring-mass-damper: GPR modelling results for the noise variance of the associated ARX models.

	GP-ARX$_1$(38,38)	GP-ARX$_3$(38,38)	GP-ARX$_5$(38,38)
fit	91.56%	83.75%	85.37%

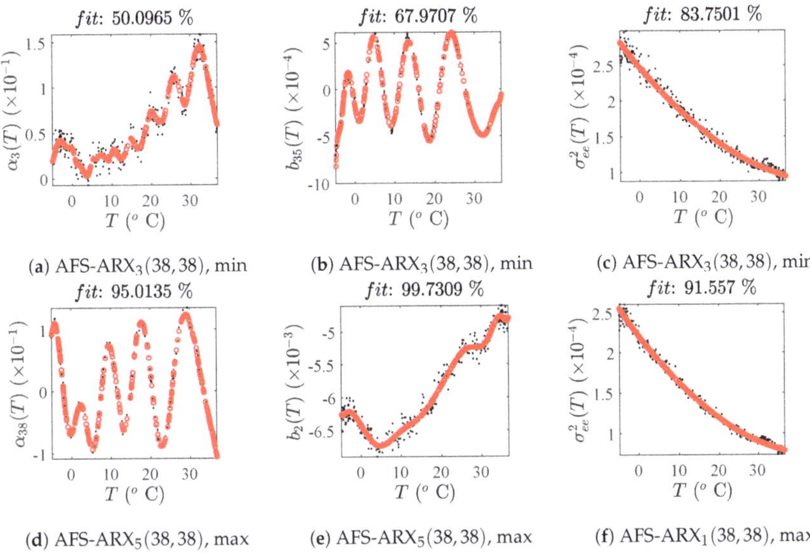

(**a**) AFS-ARX$_3$(38, 38), min (**b**) AFS-ARX$_3$(38, 38), min (**c**) AFS-ARX$_3$(38, 38), min

(**d**) AFS-ARX$_5$(38, 38), max (**e**) AFS-ARX$_5$(38, 38), max (**f**) AFS-ARX$_1$(38, 38), max

Figure 11. Spring-mass-damper: actual (black dots) and GPR-estimated (red circles) long time-scale parameter evolution.

The behaviour of the established GP-ARX models is now assessed using available information that was not used during the estimation process. In specific, the RGA stage is implemented to two long time-scale instants, which are chosen from the set $t_l = 301, \ldots, 365$. Notice that this corresponds to a form of extrapolation, since training was not applied to the whole long time-scale cycle. Figure 12 displays the RGA time-series of the available GP-ARX models for $t_l = 320$ and $t_l = 355$. A quite stable and stationary behaviour is observed, indicating that no damage has occurred within this long time-scale. Yet, in some cases the associated thresholds might result underestimated, as, for example, the ones of Figure 12d. This issue is rather attributed to the fact that the inherited uncertainties associated with the GPR models (e.g., the GPR prediction intervals) are not taken into account. Despite this inconsistency, the GP-ARX models are still capable of tracking the current state of the structure, as indicated in Figure 13, where the predicted time-series are plotted over the actual ones for the RGA of Figure 12d (e.g., the one with the "worst" behaviour).

Finally, Figure 14 illustrates the behaviour of the RGA time-series under damage, during the long time-scale $t_l = 320$. The latter is introduced as a local stiffness change, which occurs abruptly at one-third of the total simulation time. The first four rows of the figure display the $r_i[k]$'s for 1% damage in K_1, K_3, K_5 and K_7, respectively, from where the detection effectiveness of the algorithm even at such low levels is demonstrated. Notice that there's no correlation of the RGA time-series amplitudes to the location of damage, since all remain around the same levels. This can be also argued for the damage severity: the RGA time-series of the last row, which corresponds to 50% damage in K_4, result with increased amplitudes (around double the ones of the first four rows), yet, again this change is quite disproportional to the size of the damage (50% over 1%).

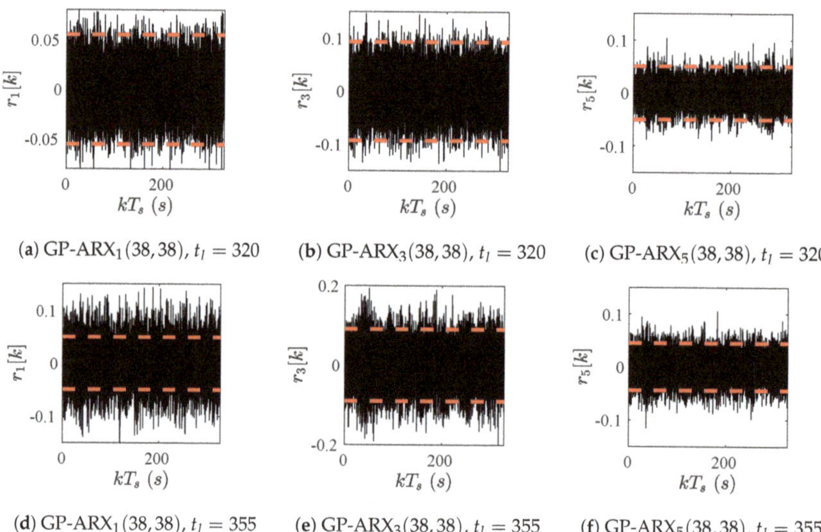

(**a**) GP-ARX$_1$(38,38), $t_l = 320$ (**b**) GP-ARX$_3$(38,38), $t_l = 320$ (**c**) GP-ARX$_5$(38,38), $t_l = 320$

(**d**) GP-ARX$_1$(38,38), $t_l = 355$ (**e**) GP-ARX$_3$(38,38), $t_l = 355$ (**f**) GP-ARX$_5$(38,38), $t_l = 355$

Figure 12. Spring-mass-damper: evolution of the residuals during two long time-scale instants from the validation set.

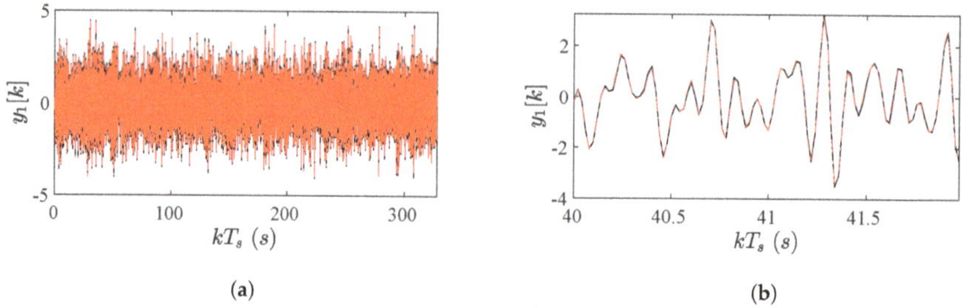

(**a**) (**b**)

Figure 13. Spring-mass-damper: actual (black curves) and predicted (dashed red curves) time-series corresponding to the RGA time-series of Figure 12d: (**a**) total time span. (**b**) ≈ 2 s zoom.

4.2. Damage Detection and Localization on a Shear Frame

The second case study pertains to the six-storey shear frame of Figure 15 under base (e.g., earthquake) excitation. It is again assumed that the stiffness is exponentially depended on temperature via the same relationship as in the spring-mass-damper case and that the temperature varies as in Equation (36). This implies that Figures 3 and 4 are also valid here, with the only difference being that K_i is expressed now on a MN/m scale and that modal damping is kept constant at 2%. The natural frequencies of the frame are shown in Figure 16.

In deriving the theoretical form of the GP-ARX models to be estimated, the equation of motion of the frame obeys

$$\mathbf{M}\ddot{\mathbf{z}}(t) + \mathbf{D}\dot{\mathbf{z}}(t) + \mathbf{K}(T)\mathbf{z}(t) = -\mathbf{M}\mathbf{1}\ddot{x}_g(t) \tag{39}$$

where $\mathbf{z}(t)$ is the vector DOFs' displacements relative to the ground motion, e.g., $z_i = x_i - x_g$ (see Figure 15 and **1** is a column vector filled with ones. The state-space representation of Equation (39) for absolute acceleration output is

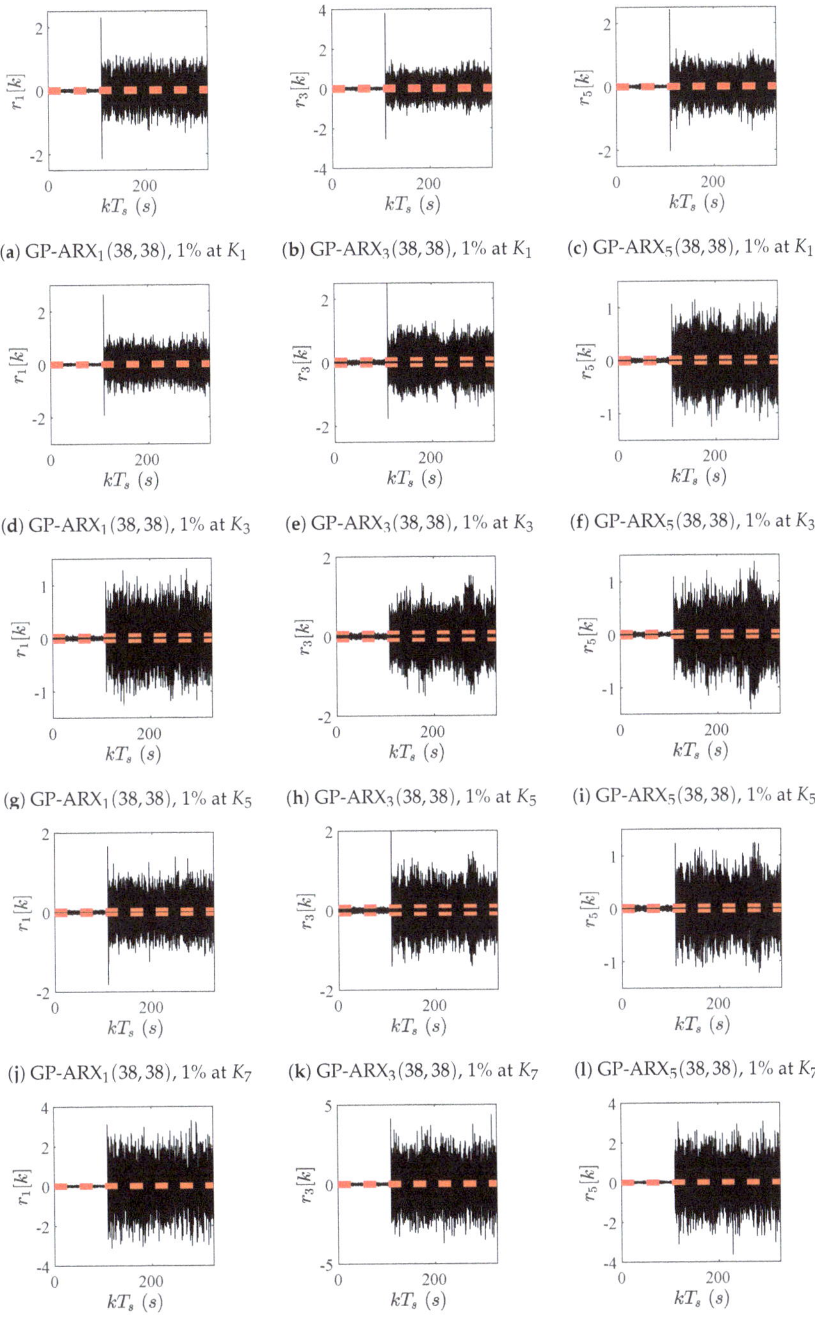

(**a**) GP-ARX$_1$(38, 38), 1% at K_1 (**b**) GP-ARX$_3$(38, 38), 1% at K_1 (**c**) GP-ARX$_5$(38, 38), 1% at K_1

(**d**) GP-ARX$_1$(38, 38), 1% at K_3 (**e**) GP-ARX$_3$(38, 38), 1% at K_3 (**f**) GP-ARX$_5$(38, 38), 1% at K_3

(**g**) GP-ARX$_1$(38, 38), 1% at K_5 (**h**) GP-ARX$_3$(38, 38), 1% at K_5 (**i**) GP-ARX$_5$(38, 38), 1% at K_5

(**j**) GP-ARX$_1$(38, 38), 1% at K_7 (**k**) GP-ARX$_3$(38, 38), 1% at K_7 (**l**) GP-ARX$_5$(38, 38), 1% at K_7

(**m**) GP-ARX$_1$(38, 38), 50% at K_4 (**n**) GP-ARX$_3$(38, 38), 50% at K_4 (**o**) GP-ARX$_5$(38, 38), 50% at K_4

Figure 14. Spring-mass-damper: RGA time-series evolution during damage at $t_l = 320$.

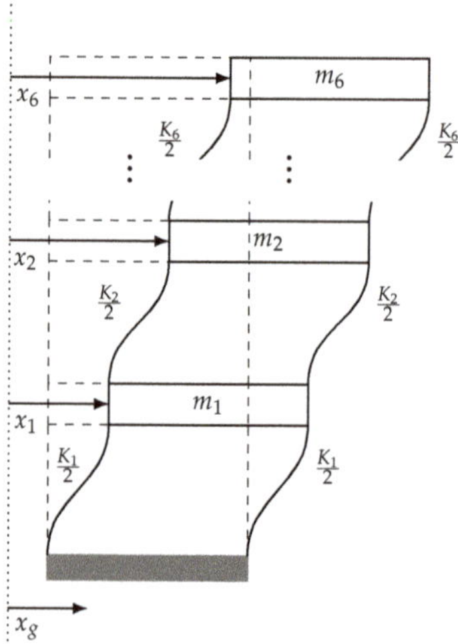

Figure 15. Six-storey shear frame with $m_i = 80$ Mgr subject to uniaxial earthquake motion.

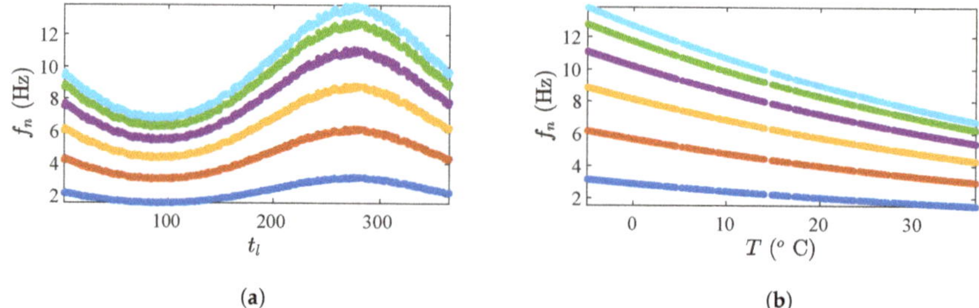

Figure 16. Shear frame: structural vibration modes for the temperature realization of Figure 3: (**a**) f_n vs. t_l. (**b**) f_n vs. T.

$$\dot{\zeta}(t) = \mathbf{A}_c(T)\zeta(t) + \mathbf{B}_c\ddot{x}_g(t) \tag{40a}$$

$$\mathbf{y}(t) = \mathbf{H}_c(T)\zeta(t) \tag{40b}$$

indicating that the equivalent digital transfer function between the acceleration of the storeys and the ground acceleration is characterized by unit delay (refer to the analysis of Section 2).

Following the steps of Section 4.1, the "healthy" structure is discretized at $T_s = 0.02$ s ($F_s = 50$ Hz) via the zero-order hold principle. Accordingly, the structural absolute vibration acceleration responses under independent realizations of zero-mean, Gaussian white noise base excitation ($\ddot{x}_g[k] \in \mathcal{N}(0, 1.87 \times 10^4)$, data length $N = 2^{14}$) are obtained and zero-mean subtracted, while it is assumed that measurements from all DOFs are available. Figure 17 shows the absolute vibration acceleration FRF of the first story at reference temperature.

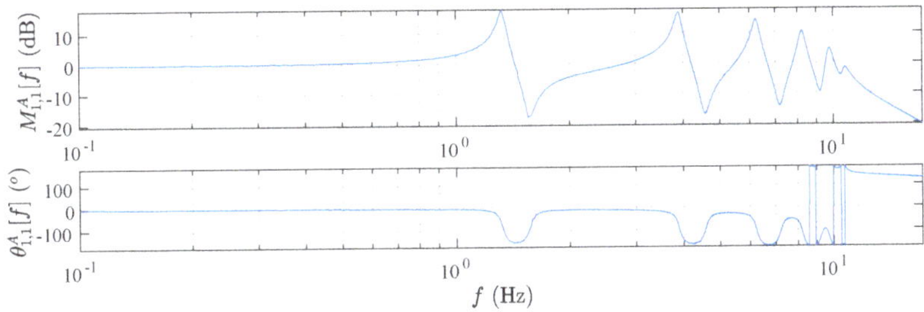

Figure 17. Shear frame: estimated (Welch's method with Hanning window, $N_{FFT} = 2^{11}$, 50% overlap) absolute vibration acceleration FRF of the first story at reference temperature.

Model order selection is based on the estimation of unit delay ARX(p, p) models for $p = 10, 12, \ldots, 50$, while using the excitation and absolute vibration acceleration response of the first storey at a randomly chosen long time instant. The performance of the model order selection criteria and the associated dispersion-corrected FSD is depicted in Figure 18, which qualify $p = 12$ as the best order. This complies with the theoretical order of Equation (15), a result that is expected due to the absence of noise-corrupted data. The behaviour of the unit delay ARX(12,12) prediction errors returns quite similar to the one of Figure 9. The subsequent identification of unit delay $\text{ARX}_{i,t_l}(12, 12)$ models for $i = 1, 2, \ldots, 6$ and $t_l = 1, \ldots, 300$ returns very high mean fitness with low dispersion, as shown in Table 4.

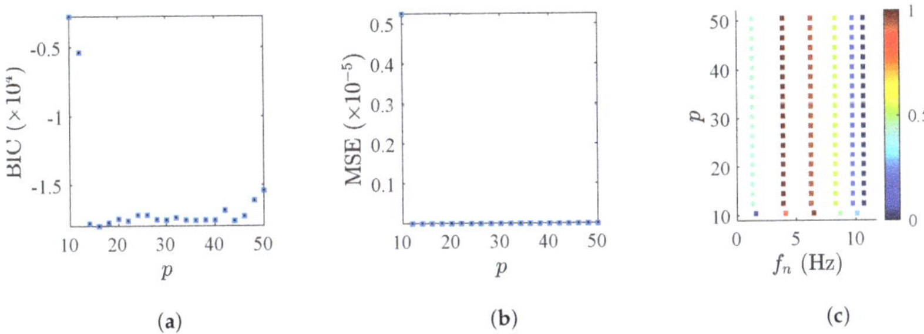

Figure 18. Shear frame: performance of the ARX model order selection criteria: (**a**) BIC. (**b**) MSE. (**c**) Dispersion-corrected frequency stabilization diagram.

Table 4. Shear frame: statistics of the percentage fitness to the short time estimation data set during ARX modelling.

	ARX_{1,t_l}(12,12)	ARX_{2,t_l}(12,12)	ARX_{3,t_l}(12,12)	ARX_{4,t_l}(12,12)	ARX_{5,t_l}(12,12)	ARX_{6,t_l}(12,12)
μ_{fit}	99.97%	99.98%	99.99%	99.98%	99.98%	98.98%
σ_{fit}	5.7×10^{-3}	4.6×10^{-3}	4.2×10^{-3}	3.9×10^{-3}	3.6×10^{-3}	3.0×10^{-3}

The GPR modelling phase is carried out using the basis functions, kernels, and orders of Table 5. Table 6 and Figure 19 show the percentage fitness, confirming the efficacy of the GPR models in fitting of the AR and eXogenous parameters. The poor results of the ARX model variances are again attributed to the noise-free data used during the identification case. The typical behaviours of the GPR models are illustrated in Figure 20.

Table 5. Shear frame: GPR basis functions, kernels, and orders.

Quantity	AR Parameters	Exogenous Parameters	Noise Variance
Basis function type	polynomial		discrete Fourier
Basis function formula	as in Table 2		as in Table 2
Kernel type	Matérn 3/2		
Kernel formula	as in Table 2		
Order	$p_a = 3$	$p_b = 3$	$p_\sigma = 16$

Table 6. Shear frame: GPR modelling results for the noise variance of the associated ARX models.

	ARX$_{1,t_l}$(12,12)	**ARX$_{2,t_l}$(12,12)**	**ARX$_{3,t_l}$(12,12)**	**ARX$_{4,t_l}$(12,12)**	**ARX$_{5,t_l}$(12,12)**	**ARX$_{6,t_l}$(12,12)**
fit	16.42%	16.49%	16.70%	16.79%	16.83%	16.74%

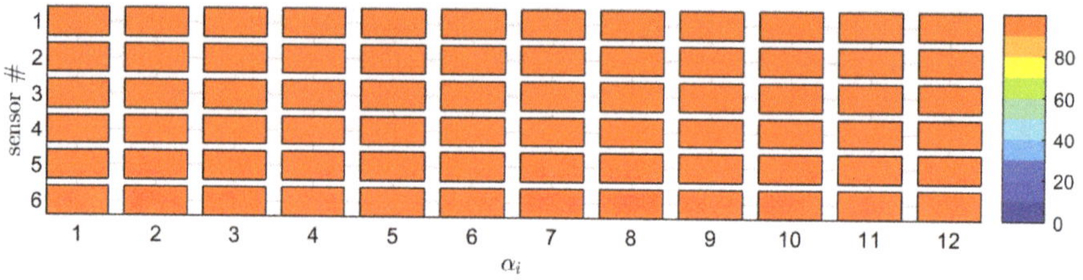

(**a**) AR parameters

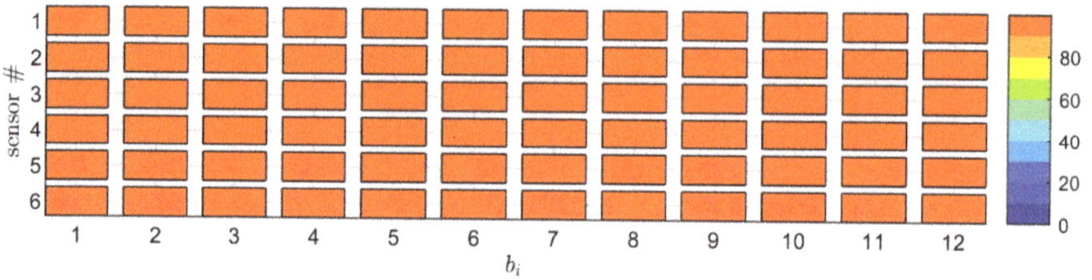

(**b**) eXogenous parameters

Figure 19. Shear frame: GPR modelling results for the AR and the eXogenous parameters of the associated ARX models.

Figure 21 displays the short time-scale temporal evolution of the RGA time-series for $t_l = 305$ under no damage. All of the residuals are stationary and within the specified thresholds, which herein result a bit overestimated, due to the low modelling quality of the respective GPR models for the ARX variances. In contrast, for the same short time-scale instant, Figure 22 illustrates the same quantities under damage, introduced as a 1% stiffness change in K_1, occurring abruptly at one-third of the total simulation time. One can here observe that there might be an indication about the location of the damage, by observing the descending dispersion from $r_1[k]$ to $r_6[k]$.

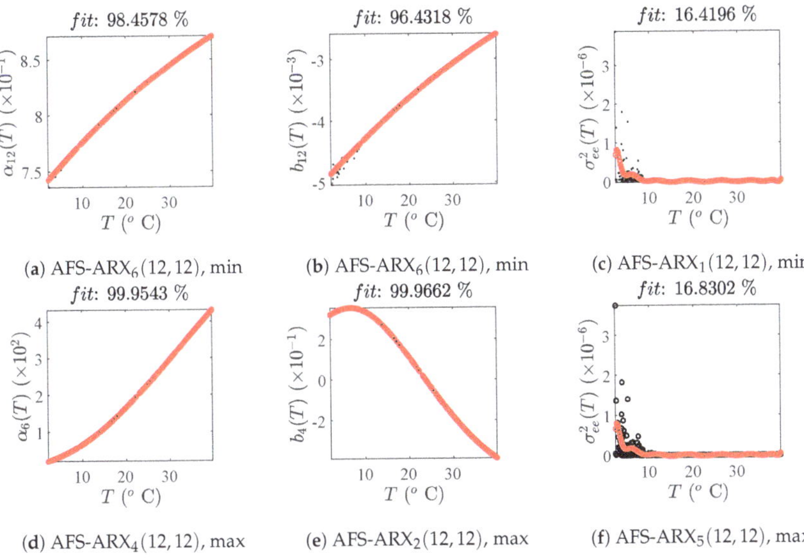

(**a**) AFS-ARX$_6$(12,12), min (**b**) AFS-ARX$_6$(12,12), min (**c**) AFS-ARX$_1$(12,12), min

(**d**) AFS-ARX$_4$(12,12), max (**e**) AFS-ARX$_2$(12,12), max (**f**) AFS-ARX$_5$(12,12), max

Figure 20. Shear frame: actual (black dots) and GPR-estimated (red circles) long time-scale parameter evolution.

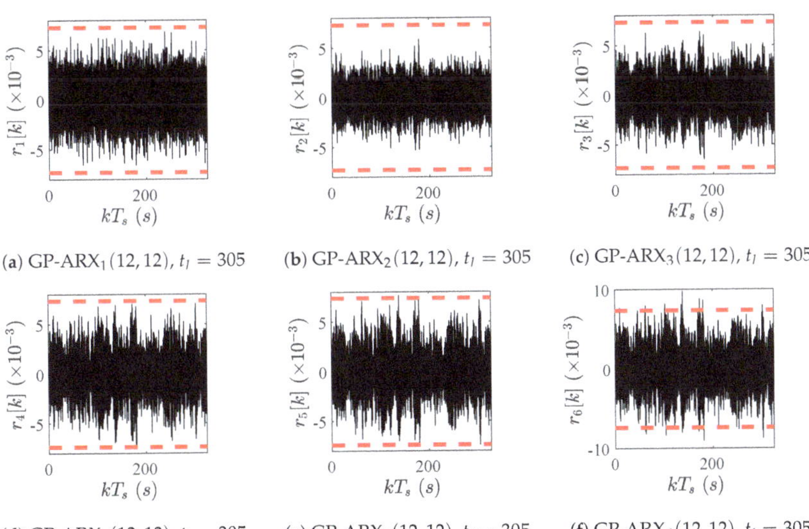

(**a**) GP-ARX$_1$(12,12), $t_l = 305$ (**b**) GP-ARX$_2$(12,12), $t_l = 305$ (**c**) GP-ARX$_3$(12,12), $t_l = 305$

(**d**) GP-ARX$_4$(12,12), $t_l = 305$ (**e**) GP-ARX$_5$(12,12), $t_l = 305$ (**f**) GP-ARX$_6$(12,12), $t_l = 305$

Figure 21. Shear frame: evolution of the residuals during a long time-scale instant from the validation set.

However, a better interpretation is offered by examining the MSCs differences, as described in Section 3.2 and Equation (35). These are plotted in Figure 23, from where it follows that the damage is indeed successfully localized in the differences of the first three modes, whereas the three trailing ones appear more "noisy" failing to localize the damage in K_1. To investigate this result, Figure 24 presents the "theoretical" MSCs differences, calculated via the eigenvalue problem of the exact structural matrices in healthy and damaged states. The close resemblance of the estimated MSCs differences to their theoretical counterparts is apparent, implying that (i) the reproduction of the modal space via the distributed GP-ARX models is to a great extend accurate; and, (ii) a damage may be localized only in lower modes.

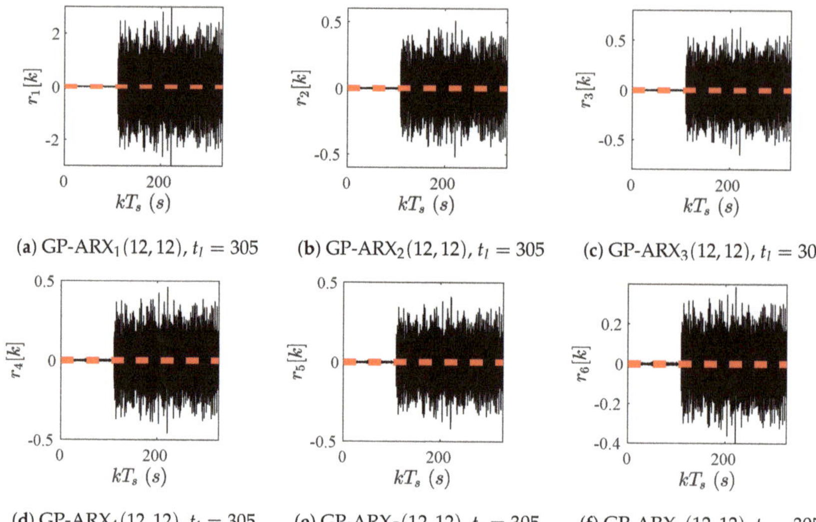

(**a**) GP-ARX$_1(12, 12)$, $t_l = 305$ (**b**) GP-ARX$_2(12, 12)$, $t_l = 305$ (**c**) GP-ARX$_3(12, 12)$, $t_l = 305$

(**d**) GP-ARX$_4(12, 12)$, $t_l = 305$ (**e**) GP-ARX$_5(12, 12)$, $t_l = 305$ (**f**) GP-ARX$_6(12, 12)$, $t_l = 305$

Figure 22. Shear frame: evolution of the residuals under 1% damage in K_1, during a long time-scale instant from the validation set.

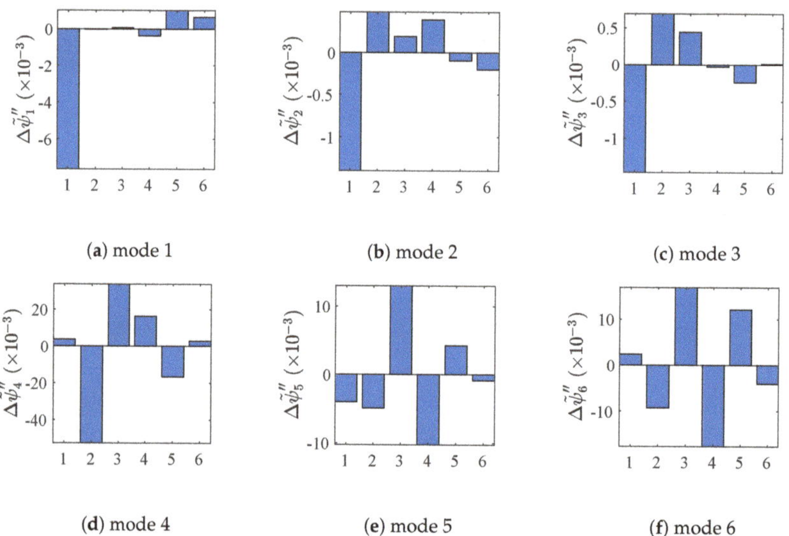

(**a**) mode 1 (**b**) mode 2 (**c**) mode 3

(**d**) mode 4 (**e**) mode 5 (**f**) mode 6

Figure 23. Shear frame: GP-ARX-based MSCs differences under 1% damage in K_1, during a long time-scale instant from the validation set.

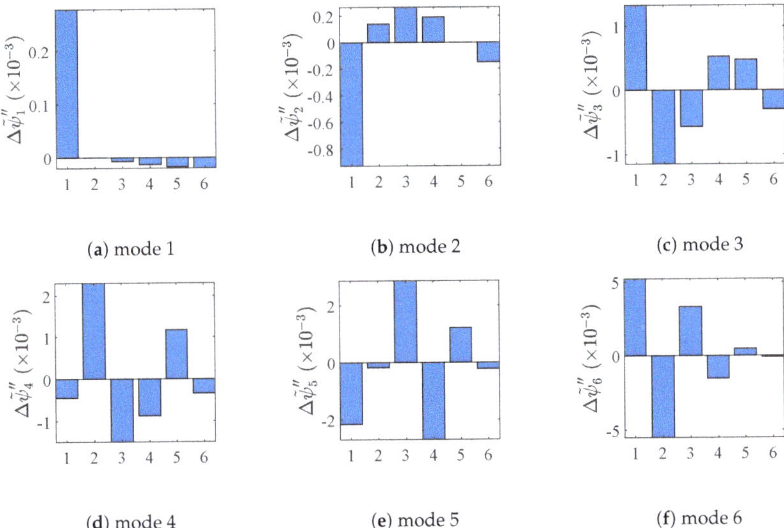

(**a**) mode 1 (**b**) mode 2 (**c**) mode 3

(**d**) mode 4 (**e**) mode 5 (**f**) mode 6

Figure 24. Shear frame: theoretical MSCs differences under 1% damage in K_1, during a long time-scale instant from the validation set.

5. Conclusions

This study tackles the damage detection and localization problem in structural systems that are amenable to environmental variations, and for which no prior information is available. The solution involves the estimation of distributed GP-ARX models for the representation of individual single input-single output pairs and the subsequent establishment of corresponding RGAs. These operate in parallel to the monitored system, are characterized by well-defined stationary statistics in the healthy state over the whole operational regime, and deviate from these in the presence of damage. In the latter case, localization is achieved by truncating all local GP-ARX models into a global state-space realization, deriving the modal space and calculating the MSC discrepancies between the healthy and damaged state-space representations.

The assessment of the method through the case studies indicates that the proposed framework is capable of detecting and localizing damage at quite low levels ("early" enough). The detection succeeds, even for a single input-output pair, while localization presumes a proper sensor network. It would be interesting to relax this requirement and attempt to integrate the damage localization task into the RGA, by configuring the residuals in some directional, or structured form; this forms current work of the authors. Nevertheless, it seems that the establishment of RGAs with localization attributes requires qualitative knowledge of structural dynamics, so that the GP-ARX parameter vector can be linked to the structural parameters.

Another issue that is observed from the case studies pertains to numerical stability. Indicatively, in the shear frame paradigm, it is found that some operating points from the validation set result in unstable GP-ARX models, even if the GPR modelling stage produces very accurate parametric representations. An alternative approach might lie in investigation of vector GP-ARX models, in order to bypass the local-to-global state-space realization, which could induce errors. Because vector models retain a considerably richer structure, an open research question would be the potential of performing GPR modelling on a reduced parameter space, similarly to what is discussed in [74].

It must be noted that structural damage may appear either as gradual degradation at slow rates (e.g., crack initiation/propagation), or as faults after "events" (e.g., an earthquake, strong wind/wave gusts, etc.). Our work is mostly set to treat this latter class: we assume distinct operational points along the long time-scale, we estimate the current state of the structure and subsequently forming/updating the RGA time-series. However, this does not imply that our method is incapable of treating the first class as well, since it has been demonstrated that the RGA can detect damage early enough, while it can easily operate in continuous mode, as long as measurements are available. However, the "footprint" of any gradual degradation on the RGA requires additional research, as mentioned above. Currently, our RGA does not correlate with damage evolution: it is just a yes/no answer to detection, which considers the long time-scale structural variability.

The encouraging results on the efficacy of the method suggest further research steps towards a more realistic SHM context. These include the addition of further free parameters into the GPR metamodeling stage, related to further environmental and operational parameters (operating points, boundary conditions, etc.), aside from temperature. The relaxation on the availability and the stationarity assumption of the input(s) forms a further critical aspect that is frequently encountered in practice: here, the investigation of scalar/vector GP-AR/ARMA models and their ARIMA counterparts form a straightforward task, which is not yet adequately treated in existing literature. Finally, an experimental validation is currently underway by the authors.

Author Contributions: Conceptualization, K.T., V.D., Y.O. and E.C.; methodology, K.T., V.D. and Y.O.; software, K.T. and V.D.; validation, K.T., V.D. and Y.O.; writing—original draft preparation, K.T. and V.D.; writing—review and editing, V.D and E.C.; supervision, V.D. and E.C.; project administration, E.C.; funding acquisition, E.C. All authors have read and agreed to the published version of the manuscript.

Funding: This research was funded by the European Research Council via the ERC Starting Grant WINDMIL (ERC-2015-StG #679843) on the topic of "Smart Monitoring, Inspection and Life-Cycle Assessment of Wind Turbines", as well as the ERC Proof of Concept (PoC) Grant, ERC-2018-PoC WINDMIL RT-DT on "An autonomous Real-Time Decision Tree framework for monitoring and diagnostics on wind turbines".

Conflicts of Interest: The authors declare no conflict of interest.

References

1. Ding, S. *Model-Based Fault Diagnosis Techniques: Design Schemes, Algorithms and Tools*; Springer: London, UK, 2013.
2. Gertler, J. *Fault Detection and Diagnosis in Engineering Systems*; Marcel–Dekker, Inc.: New York, NY, USA, 1998.
3. Natke, H.; Cempel, C. *Model—Based Fault Diagnosis of Mechanical Systems: Fundamentals, Detection, Localization, Assessment*; Springer: Berlin, Germany, 1997.
4. Chen, J.; Patton, R. *Robust Model—Based Fault Diagnosis of Dynamic Systems*; Kluwer Academic Publishers: Norwell, MA, USA, 1999.
5. Kullaa, J. Development of virtual sensors to increase the sensitivity to damage. *Procedia Eng.* **2017**, *199*, 1937–1942. [CrossRef]
6. Maes, K.; Lombaert, G. Fatigue monitoring of railway bridges by means of virtual sensing. In Proceedings of the Belgian and Dutch National Groups of IABSE—Young Engineers Colloquium, TU/e, Eindhoven, The Netherlands, 15–16 March 2019.
7. Tatsis, K.; Dertimanis, V.; Abdallah, I.; Chatzi, E. A substructure approach for fatigue assessment on wind turbine support structures using output-only measurements. *Procedia Eng.* **2017**, *199*, 1044–1049. [CrossRef]
8. Tatsis, K.; Dertimanis, V.; Chatzi, E. Response Prediction of Systems Experiencing Operational and Environmental Variability. In *Computing in Civil Engineering 2019*; Chapter Computing in Civil Engineering 2019: Smart Cities, Sustainability, and Resilience; ASCE: Reston, VA, USA, 2019; pp. 468–474. [CrossRef]
9. Vettori, S.; Di Lorenzo, E.; Cumbo, R.; Musella, U.; Tamarozzi, T.; Peeters, B.; Chatzi, E. Kalman-based Virtual Sensing for Improvement of Service Responses Replication in Environmental Tests #8071. In Proceedings of the IMAC XXXVIII Conference, Orlando, FL, USA, 28–31 January 2019.
10. Lourens, E.; Papadimitriou, C.; Gillijns, S.; Reynders, E.; Roeck, G.D.; Lombaert, G. Joint input-response estimation for structural systems based on reduced-order models and vibration data from a limited number of sensors. *Mech. Syst. Signal Proces.* **2012**, *29*, 310–327. [CrossRef]

11. Sedehi, O.; Papadimitriou, C.; Teymouri, D.; Katafygiotis, L.S. Sequential Bayesian estimation of state and input in dynamical systems using output-only measurements. *Mech. Syst. Signal Proces.* **2019**, *131*, 659–688. [CrossRef]
12. Ou, Y.; Chatzi, E.N.; Dertimanis, V.K.; Spiridonakos, M.D. Vibration-based experimental damage detection of a small-scale wind turbine blade. *Struct. Health Monit.* **2017**, *16*, 79–96. [CrossRef]
13. Xiang, J.; Lei, Y.; Wang, Y.; He, Y.; Zheng, C.; Gao, H. Structural Dynamical Monitoring and Fault Diagnosis. *Shock Vib.* **2015**, *2015*, 193831. [CrossRef]
14. Wu, J.; Rudnick, E.M. Bridge fault diagnosis using stuck-at fault simulation. *IEEE Trans. Comput. Aided Des. Integr. Circuits Syst.* **2000**, *19*, 489–495. [CrossRef]
15. Xiang, T.; Huang, K.; Zhang, H.; Zhang, Y.; Zhang, Y.; Zhou, Y. Detection of Moving Load on Pavement Using Piezoelectric Sensors. *Sensors* **2020**, *20*, 2366. [CrossRef]
16. Alblawi, A. Fault diagnosis of an industrial gas turbine based on the thermodynamic model coupled with a multi feedforward artificial neural networks. *Energy Rep.* **2020**, *6*, 1083–1096. [CrossRef]
17. Kazemi, H.; Yazdizadeh, A. Fault detection and isolation of gas turbine engine using inversion-based and optimal state observers. *Eur. J. Control* **2020**. [CrossRef]
18. Abdallah, I.; Dertimanis, V.; Mylonas, H.; Tatsis, K.; Chatzi, E.; Dervilis, N.; Worden, K.; Maguire, E. Fault Diagnosis of wind turbine structures using decision tree learning algorithms with big data. In Proceedings of the European Safety and Reliability Conference, ESREL 2018, Trondheim, Norway, 17–21 June 2018.
19. Sanchez, H.; Escobet, T.; Puig, V.; Odgaard, P. Fault diagnosis of an advanced wind turbine benchmark using interval-based ARRs and observers. *IEEE Trans. Ind. Electr.* **2015**, *62*, 3783–3793. [CrossRef]
20. Badihi, H.; Zhang, Y.; Hong, H. Wind Turbine Fault Diagnosis and Fault-Tolerant Torque Load Control Against Actuator Faults. *IEEE Trans. Control Syst. Technol.* **2015**, *23*, 1351–1372. [CrossRef]
21. Cross, E.J.; Worden, K.; Farrar, C.R. Structural Health Monitoring for Civil Infrastructure. In *Health Assessment of Engineered Structures: Bridges, Buildings and Other Infrastructures*; Haldar, A., Ed.; World Scientific: Singapore, 2013; Chapter 1.
22. Tcherniak, D.; Mølgaard, L.L. Vibration-based SHM System: Application to Wind Turbine Blades. *J. Phys. Conf. Ser.* **2015**, *628*, 012072. [CrossRef]
23. Ou, Y.; Grauvogl, B.; Spiridonakos, M.; Dertimanis, V.; Chatzi, E.; Vidal, J. Vibration-based Damage Detection on a Blade of a Small Scale Wind Turbine. *Proc. IWSHM* **2015**. [CrossRef]
24. Dutta, A.; McKay, M.E.; Kopsaftopoulos, F.; Gandhi, F. Fault Detection and Identification for Multirotor Aircraft by Data-Driven Statistical Learning Methods. In Proceedings of the AIAA Propulsion and Energy 2019 Forum, Indianapolis, IN, USA, 19–22 August 2019. [CrossRef]
25. Marzat, J.; Piet-Lahanier, H.; Damongeot, F.; Walter, E. Model-based fault diagnosis for aerospace systems: a survey. *Proc. Inst. Mech. Eng. Part G J. Aerosp. Eng.* **2012**, *226*, 1329–1360. [CrossRef]
26. Giurgiutiu, V.; Zagrai, A.N. Electro-mechanical impedance method for crack detection in metallic plates. In *Advanced Nondestructive Evaluation for Structural and Biological Health Monitoring*; Kundu, T., Ed.; International Society for Optics and Photonics, SPIE: Washington, DC, USA, 2001; Volume 4335, pp. 131–142.
27. Tashakori, S.; Baghalian, A.; Cuervo, J.; Senyurek, V.Y.; Tansel, I.N.; Uragun, B. Inspection of the machined features created at the embedded sensor aluminum plates. In Proceedings of the 2017 8th International Conference on Recent Advances in Space Technologies (RAST), Istanbul, Turkey, 19–22 June 2017; pp. 517–522.
28. Xu, B.; Giurgiutiu, V. A Low-Cost and Field Portable Electromechanical (E/M) Impedance Analyzer for Active Structural Health Monitoring. In Proceedings of the 5th International Workshop on Structural Health Monitoring, Stanford University, Stanford, CA, USA, 15–17 September 2005.
29. Sohn, H. Effects of environmental and operational variability on structural health monitoring. *Philos. Trans. Ser. A Math. Phys. Eng. Sci.* **2007**, *365*, 539–560. [CrossRef] [PubMed]
30. Kullaa, J. Distinguishing between sensor fault, structural damage, and environmental or operational effects in structural health monitoring. *Mech. Syst. Signal Proces.* **2011**, *25*, 2976–2989. [CrossRef]
31. Cross, E.; Koo, K.; Brownjohn, J.; Worden, K. Long-term monitoring and data analysis of the Tamar Bridge. *Mech. Syst. Signal Proces.* **2013**, *35*, 16–34. [CrossRef]
32. Cornwell, P.; Farrar, C.; Doebling, S.; Sohn, H. Environmental variability of modal properties. *Exp. Tech.* **1999**, *23*, 45–48. [CrossRef]

33. Liu, C.; De Wolf, J. Effect of temperature on modal variability of a curved concrete bridge under ambient loads. *J. Struct. Eng.* **2007**, *133*, 1742–1751. [CrossRef]
34. Rohrmann, R.; Baessler, M.; Said, S.; Schmid, W.; Ruecker, W. Structural causes of temperature affected modal data of civil structures obtained by long time monitoring. In Proceedings of IMAC XVIII—18th International Modal Analysis Conference, San Antonio, TX, USA, 7–10 February 2000.
35. Bernal, D. Analytical techniques for damage detection and localization for assessing and monitoring civil infrastructures. In *Sensor Technologies for Civil Infrastructures*; Wang, M., Lynch, J., Sohn, H., Eds.; Woodhead Publishing: Cambridge, UK, 2014; Chapter 3; Volume 2, pp. 67–92.
36. Figueiredo, E.; Park, G.; Farrar, C.; Worden, K.; Figueiras, J. Machine learning algorithms for damage detection under operational and environmental variability. *Struct. Health Monit.* **2011**, *10*, 559–572. [CrossRef]
37. Harmanci, Y.; Spiridonakos, M.; Chatzi, E.; Kübler, W. An Autonomous Strain-Based Structural Monitoring Framework for Life-Cycle Analysis of a Novel Structure. *Front. Built Environ.* **2016**, *2*, 13. [CrossRef]
38. Sohn, H.; Worden, K.; Farrar, C. Novelty detection under changing environmental conditions. In Proceedings of the SPIE4330: 8th Annual International Symposium on Smart Structures and Materials, Newport Beach, CA, USA, 4–8 March 2001; pp. 108–118.
39. Dervilis, N.; Choi, M.; Taylor, S.; Barthorpe, R.; Park, G.; Farrar, C.; Worden, K. On damage diagnosis for a wind turbine blade using pattern recognition. *J. Sound Vib.* **2014**, *333*, 1833–1850. [CrossRef]
40. Lämsä, V.; Kullaa, J. Nonlinear Factor Analysis in Structural Health Monitoring to Remove Environmental Effects. In Proceedings of the 6th International Workshop on Structural Health Monitoring, Stanford University, Stanford, CA, USA, 11–13 September 2007.
41. Kullaa, J. Elimination of environmental influences from damage-sensitive features in a structural health monitoring system. In Proceedings of the First European Workshop on Structural Health Monitoring, Paris, France, 10–12 July 2002; pp. 742–749.
42. Manson, G. Identifying damage sensitive, environment insensitive features for damage detection. In Proceedings of the 3rd International Conference on Identification in Engineering Systems, Swansea, UK, 15–17 April 2002; pp. 187–197.
43. Reynders, E.; Wursten, G.; De Roeck, G. Output-only structural health monitoring in changing environmental conditions by means of nonlinear system identification. *Struct. Health Monit.* **2013**. [CrossRef]
44. De Boe, P.; Golinval, J. Principal component analysis of a piezosensor array for damage localization. *Struct. Health Monit.* **2003**, *2*, 137–144. [CrossRef]
45. García, D.; Trendafilova, I. A multivariate data analysis approach towards vibration analysis and vibration-based damage assessment: Application for delamination detection in a composite beam. *J. Sound Vib.* **2014**, *333*, 7036–7050. [CrossRef]
46. Shokrani, Y.; Dertimanis, V.K.; Chatzi, E.N.; Savoia, M.N. On the use of mode shape curvatures for damage localization under varying environmental conditions. *Struct. Control Health Monit.* **2018**, *25*, e2132. [CrossRef]
47. Ou, Y.W.; Dertimanis, V.K.; Chatzi, E.N. Operational Damage Localization of Wind Turbine Blades. In *Experimental Vibration Analysis for Civil Structures*; Conte, J.P., Astroza, R., Benzoni, G., Feltrin, G., Loh, K.J., Moaveni, B., Eds.; Springer International Publishing: Cham, Switzerland, 2018; pp. 261–272.
48. Dervilis, N.; Worden, K.; Cross, E. On robust regression analysis as a means of exploring environmental and operational conditions for SHM data. *J. Sound Vib.* **2015**, *347*, 279–296. [CrossRef]
49. Dervilis, N.; Zhang, T.; Bull, L.; Cross, E.; Rogers, T.; Fuentes, R.; Dertimanis, V.; Abdallah, I.; Chatzi, E.; Worden, K. A nonlinear robust outlier detection approach for SHM. In Proceedings of the IOMAC 2019, Copenhagen, Denmark, 13–15 May 2019.
50. Toth, R. *Modeling and Identification of Linear Parameter-Varying Systems*; Lecture Notes in Control and Information Sciences; Springer: Berlin/Heidelberg, Germany, 2010. [CrossRef]
51. Avendaño-Valencia, L.D.; Chatzi, E.N.; Koo, K.Y.; Brownjohn, J.M. Gaussian Process Time-Series Models for Structures under Operational Variability. *Front. Built Environ.* **2017**, *3*, 69. [CrossRef]
52. Kopsaftopoulos, F.; Nardari, R.; Li, Y.H.; Chang, F.K. A stochastic global identification framework for aerospace structures operating under varying flight states. *Mech. Syst. Signal Proces.* **2018**, *98*, 425–447. [CrossRef]
53. Sakellariou, J.; Fassois, S. Functionally Pooled models for the global identification of stochastic systems under different pseudo-static operating conditions. *Mech. Syst. Signal Proces.* **2016**, *72–73*, 785–807. [CrossRef]

54. Spiridonakos, M.D.; Chatzi, E.N.; Sudret, B. Polynomial Chaos Expansion Models for the Monitoring of Structures under Operational Variability. *ASCE-ASME J. Risk Uncertain. Eng. Syst. Part A Civ. Eng.* **2016**, *2*, B4016003. [CrossRef]
55. Spiridonakos, M.D.; Chatzi, E.N. Stochastic Structural Identification from Vibrational and Environmental Data. In *Encyclopedia of Earthquake Engineering*; Beer, M., Kougioumtzoglou, I.A., Patelli, E., Au, I.S.K., Eds.; Springer: Berlin/Heidelberg, Germany, 2014; pp. 1–16. [CrossRef]
56. Avendaño-Valencia, L.D.; Fassois, S.D. Natural vibration response based damage detection for an operating wind turbine via Random Coefficient Linear Parameter Varying AR modelling. *J. Phys. Conf. Ser.* **2015**, *628*, 012073. [CrossRef]
57. Bogoevska, S.; Spiridonakos, M.; Chatzi, E.; Dumova-Jovanoska, E.; Höffer, R. A Data-Driven Diagnostic Framework for Wind Turbine Structures: A Holistic Approach. *Sensors* **2017**, *17*, 720. [CrossRef]
58. Avendaño-Valencia, L.D.; Fassois, S.D. Gaussian Mixture Random Coefficient model based framework for SHM in structures with time–dependent dynamics under uncertainty. *Mech. Syst. Signal Proces.* **2017**, *97*, 59–83. [CrossRef]
59. Avendaño-Valencia, L.D.; Fassois, S.D. Damage/fault diagnosis in an operating wind turbine under uncertainty via a vibration response Gaussian mixture random coefficient model based framework. *Mech. Syst. Signal Proces.* **2017**, *91*, 326–353. [CrossRef]
60. Avendaño-Valencia, L.D.; Chatzi, E.N. Sensitivity driven robust vibration-based damage diagnosis under uncertainty through hierarchical Bayes time-series representations. *Procedia Eng.* **2017**, *199*, 1852–1857. [CrossRef]
61. Demetriou, M.A. Using unknown input observers for robust adaptive fault detection in vector second-order systems. *Mech. Syst. Signal Proces.* **2005**, *19*, 291–309. [CrossRef]
62. Basseville, M. Information criteria for residual generation and fault detection and isolation. *Automatica* **1997**, *33*, 783–803. [CrossRef]
63. Isermann, R. *Fault-Diagnosis Systems: An Introduction from Fault Detection to Fault Tolerance*; Springer: Berlin, Germany, 2006.
64. Simani, S.; Fantuzzi, C.; Patton, R. *Model-Based Fault Diagnosis in Dynamic Systems Using Identification Techniques*; Springer: London, UK, 2003.
65. Dertimanis, V.; Giagopoulos, D.; Chatzi, E. Finite Element Metamodelling of Uncertain Structures. In Proceedings of the ECCOMAS 2016, Crete, Greece, 5–10 June 2016; pp. 6316–6327.
66. Spiridonakos, M.; Dertimanis, V.; Chatzi, E. Estimation of Data-Driven Polynomial Chaos using Hybrid Evolution Strategies. In *Proceedings of the Fourth International Conference on Soft Computing Technology in Civil, Structural and Environmental Engineering;* Tsompanakis, Y., Kruis, J., Topping, B., Eds.; Civil-Comp Press: Stirlingshire, UK, 2015; p. 24. [CrossRef]
67. Dertimanis, V.; Spiridonakos, M.; Chatzi, E. Data-driven uncertainty quantification of structural systems via B-spline expansion. *Comput. Struct.* **2018**, *207*, 245–257. [CrossRef]
68. Box, G.; Jenkins, G.; Reinsel, G. *Time Series Analysis, Forecasting and Control*, 4th ed.; John Wiley & Sons Ltd.: Chichester, UK, 2013.
69. Dertimanis, V.K.; Koulocheris, D.V. VAR Based State–space Structures: Realization, Statistics and Spectral Analysis. In *Lecture Notes in Electrical Engineering*; Cetto, J.A., Filipe, J., Ferrier, J.L., Eds.; Springer: Berlin/Heidelberg, Germany, 2011; Volume 85, pp. 301–315.
70. Dertimanis, V.K. On the use of dispersion analysis for model assessment in structural identification. *J. Vib. Control* **2013**, *19*, 2270–2284. [CrossRef]
71. Dertimanis, V.K.; Chatzi, E. Dispersion-corrected stabilization diagrams for model order assessment in structural identification. In Proceedings of the 7th European Workshop on Structural Health Monitoring, EWSHM 2014, Nantes, France, 8–11 July 2014; pp. 2159–2166.
72. Dertimanis, V.K.; Spiridonakos, M.D.; Chatzi, E.N. Dispersion–Corrected, Operationally Normalized Stabilization Diagrams for Robust Structural Identification. In *Model Validation and Uncertainty Quantification;* Atamturktur, H.S., Moaveni, B., Papadimitriou, C., Schoenherr, T., Eds.; Springer International Publishing: Cham, Switzerland, 2015; Volume 3, pp. 75–83. [CrossRef]

73. Pandey, A.; Biswas, M.; Samman, M. Damage detection from changes in curvature mode shapes. *J. Sound Vib.* **1991**, *145*, 321–332. [CrossRef]
74. Avendaño-Valencia, L.D.; Chatzi, E.N. Multivariate GP-VAR models for robust structural identification under operational variability. *Probabilistic Eng. Mech.* **2020**, *60*, 103035. [CrossRef]

MDPI
St. Alban-Anlage 66
4052 Basel
Switzerland
Tel. +41 61 683 77 34
Fax +41 61 302 89 18
www.mdpi.com

Journal of Sensor and Actuator Networks Editorial Office
E-mail: jsan@mdpi.com
www.mdpi.com/journal/jsan

www.ingramcontent.com/pod-product-compliance
Lightning Source LLC
LaVergne TN
LVHW070843170726
843515LV00005B/561